P9-BBO-356

THE BOY WHO PLAYED WITH FUSION

THE BOY WHO PLAYED WITH FUSION

Extreme Science, Extreme Parenting, and How to Make a Star

Tom Clynes

An Eamon Dolan Book
Houghton Mifflin Harcourt
BOSTON NEW YORK
2015

www.hmhco.com

Library of Congress Cataloging-in-Publication Data
Clynes, Tom, author.
The boy who played with fusion : extreme science, extreme parenting,
and how to make a star / Tom Clynes.
pages cm
"An Eamon Dolan book."
Includes bibliographical references and index.
ISBN 978-0-544-08511-4 (hardcover) — ISBN 978-0-544-08474-2 (ebook)
1. Wilson, Taylor, 1994– 2. Gifted boys—United States—Biography.
3. Fusion reactors. 4. Nuclear fusion. I. Title.
QC774.W55C59 2015
539.7'64092 — dc23
[B]
2014048519

Book design by Chrissy Kurpeski
Typeset in Minion Pro

Printed in the United States of America
1 2 3 4 5 6 7 8 9 10 DOC

Portions of this book originally appeared in slightly different form in *Popular Science*.

To my sons, Charlie and Joe

CONTENTS

PART IV

PART V

Penetrating so many secrets, we cease to believe in the unknowable. But there it sits nevertheless, calmly licking its chops.

— H. L. MENCKEN

Add to this cruelly delicate organism the overpowering necessity to create, create, create — so that without the creating of music or poetry or books or buildings or something of meaning, his very breath is cut off from him. He must create, must pour out creation. By some strange, unknown, inward urgency he is not really alive unless he is creating.

— PEARL S. BUCK

Hi, my name is Taylor Wilson and I am 15 years old. I am an obsessive lover of all things nuclear and have a home amateur nuclear laboratory.

— TAYLOR'S NUKE SITE HOMEPAGE

INTRODUCTION

"PROPULSION," THE NINE-YEAR-OLD says as he leads his dad through the gate of the U.S. Space and Rocket Center. "I just want to see the propulsion stuff."

Situated next to the northern Alabama complex where NASA engineers designed and built the rockets that propelled America's space program, the center showcases the world's most impressive collection of high-flying hardware. Visitors can touch the scorched Apollo 16 command module, tumble-spin in a chair that mimics the frictionless vacuum of space, or command a mission in the space shuttle's cockpit simulator.

But Taylor Wilson mostly wants to see the museum's prize piece, the massive Saturn V rocket that launched mankind to the moon. Late that afternoon, father and son step inside the building built to house the reclining rocket, restored to its Apollo-era readiness. The tour guide, a young woman, leads their group of parents and children along the 363-foot-long behemoth suspended above the floor. As they duck under its five exhaust nozzles, each a dozen feet across, Kenneth Wilson glances at his awestruck boy and feels his burden beginning to lift. For a few minutes, at least, someone else will feed his son's relentless appetite for knowledge.

The docent tells the group that the Saturn V is the most powerful rocket ever built. Churning out a million and a half pounds of thrust, it boosted America decisively past the Soviets, spiriting two dozen Apollo astronauts to the moon. Though the three-stage rocket is retired now, the guide says, it remains unmatched in its capacity to lift men and gear beyond the tug of Earth's gravity. Producing thirty-two million horsepower at full blast, the Saturn could accelerate a spaceship from zero to seventeen thousand miles per hour in eight minutes.

Then Taylor raises his hand—not with a question, but an answer. He knows what makes this rocket go up. And he wants—he needs—to tell everyone about it, about how acceleration relates to exhaust velocity and dynamic mass, about payload ratios, about the mix of kerosene and liquid oxygen that the first stage burned at six thousand pounds per second. The guide takes a step back, yielding the floor to this slender, overexcited kid who's unleashing a torrent of PhD-level concepts in a deep-Arkansas drawl as if there weren't enough seconds in a day for him to blurt it all out. The other adults step back too, perhaps jolted off balance by the incongruities of age and audacity, intelligence and exuberance.

The nine-year-old tells the group that he designs and builds his own rockets at home in Texarkana. Next, he's talking about the Saturn's solid-fuel second and third stages, the comparative advantages of their propellants, the tradeoffs rocket designers make between thrust and cost, weight and safety. As the guide runs to fetch her boss — *You gotta see this kid!* — Kenneth feels the weight coming down on him again.

He and his wife, Tiffany, have done everything they can to nourish their older son's manic, metastasizing curiosity. Since the first moments of his existence, Taylor has complicated, confounded, and chaoticized nearly every detail of his family's lives. Indeed, Kenneth will look back on this particular day as one of the uncomplicated ones, when his scary-smart son was into relatively simple things, like rocket science.

This was before Taylor transformed the family's garage into a trove of glowing rocks and liquids and metals with enigmatic and terrifying powers . . . before he built a reactor that could hurl atoms together in a 500-million-degree plasma core, becoming, at fourteen, the youngest person on Earth to achieve nuclear fusion . . . before the creations of his restless intellect astounded everyone from the president of the United States to the audiences at TED Talks . . . before he conceived, in a series of unlikely epiphanies, new ways to use subatomic particles to confront some of the biggest challenges of our time: cancer, nuclear terrorism, sustainable energy.

This book had its beginnings in 2010 when, as a contributing editor at *Popular Science* magazine, I discovered a small community of nuclear physics enthusiasts, high-energy hobbyists who were taking on both the formidable theory and the precision engineering of applied nuclear science. The idea that self-taught amateurs outside the Big Science world of billion-dollar research laboratories were tinkering with nukes — fusing atomic nuclei, transmuting elements, constructing atom-smashing machines in self-built laboratories—was both intriguing and unsettling. Members of this guarded clique began to open up to me, and one of them mentioned a fourteen-year-old boy from Texarkana who had just become one of only thirty-two individuals on the planet to build a working nuclear fusion reactor, a miniature sun on Earth.

And yet, what would set Taylor apart was not his machine or his intellect but his buoyantly audacious approach to science, and life. I'd met a few child geniuses, and I could tell immediately that Taylor's genius was a different kind.

His is not the eyes-down, inwardly focused kind that skulks in the corner at the science fair. Nor is it the socially maladroit kind typified by Sheldon Cooper, the theoretical physicist in the television series *The Big Bang Theory.* Taylor's genius is eyes-up and hands-on, and exuberantly connected to the universe. Indeed, it is his gift for creating connections—personal, intellectual, practical—that has allowed him to build a world for himself that seems to have few limits.

"Within two minutes of meeting him," says Stephen Younger, the former head of nuclear weapons research at Los Alamos National Laboratory, "you realize that the kinds of things that most people think are impossible, Taylor just goes out and does."

You also realize that, despite his precocity and the Einsteinian zeal of his curiosity, Taylor is in many ways a normal kid with a normal (though often baffled) circle of family and friends, a normal teenager's series of crushes and confusions, and a still-developing identity. At first a timid child, he would burgeon into a garrulous, science-obsessed grade-schooler with a passion for explosive chemistry that would progress to an obsessive need to understand the mysteries of the subatomic world. At the age of eleven, distraught over his grandmother's impending death, Taylor would experience an illuminating moment in which he'd envision not only a solution that could help millions but also an image of his future self, transformed from curious child to groundbreaking nuclear physicist. The clarity of his vision, and his belief that he could achieve it, would open new galaxies of possibilities for Taylor and sustain and motivate him as he pursued his unlikely dream.

The attention that the *Popular Science* feature generated caught me offguard. I'd imagined that people would find Taylor's story fascinating, but I hadn't anticipated how deeply readers would connect with it emotionally. Beyond the newsworthiness of a fourteen-year-old achieving nuclear fusion, many people said they were inspired by the sense of wonder and surprise—and, especially, optimism—that had originally drawn me into Taylor's world.

For me, the story was a departure from the kinds of things I had spent much of my career covering: subjects like Ebola epidemics, eco-mercenaries, and career-ending scientific slap-fights. For the most part, I don't *do* optimistic. But I found Taylor's story—or maybe it found me—at the right moment, during a bleak period in my own life, just after my ten-year marriage had fallen apart. The hopefulness at the heart of Taylor's driving need to understand the world and make it better helped me reimagine the possibilities ahead for myself and my own young children, and for the world they would grow into.

As a parent, I was inspired by Kenneth and Tiffany's often counterintuitive approach to nurturing their children's talents. The lengths to which they—and the educators and mentors they brought into Taylor's orbit—were willing to go to support their son as he pursued his unnerving interests were, to me, even more impressive than Taylor's intrinsic talents. I would come to understand that what Taylor would achieve was a product of not only his intellectual gifts, but of the fact that he'd been gifted with parents of the most extraordinary sort.

Taylor, with his nonscientist parents and his early upbringing outside the elite zones of education, didn't emerge from what we often consider the typical wellsprings of prodigiousness. Child-development experts, educators, neuroscientists, and cognitive psychologists are just now beginning to understand the complex mix of genetics and environment that creates a child like Taylor Wilson. At this point, we don't know when or where a prodigy will pop out of the population. We do, however, know how to spot one.

Four decades of tracking data have made it clear that many of the innovators who are transforming society, advancing knowledge, and reinventing culture are in the top 1 percent in intellectual ability—and that many of them were identified as top performers by their teenage years. For instance, Mark Zuckerberg and Sergey Brin each attended a summer program sponsored by the Center for Talented Youth, open, at the time, to kids who scored in the top 1 percent on standardized tests. Bill Gates was in the top 1 percent; Steve Jobs too. Many high-achieving nontechies are as well: Stefani Germanotta, a.k.a. Lady Gaga, was enrolled in the same program as Zuckerberg and Brin.

Unfortunately, of the millions of potential world-changers who are born each year, only a small portion will be noticed and given the resources they need to develop their prodigious gifts. By refusing them an education that is appropriate to their abilities, we are potentially hobbling our economies and denying our civilization its next generation of innovators—the Salks, Mozarts, and Curies who can push the frontiers of knowledge forward.

Education researchers now estimate that the academically gifted make up 6 to 10 percent of the U.S. school-age population. When the definition of *gifted* is expanded to include artistic, athletic, and other talents, the proportion is much higher. In fact, the latest research suggests that nearly everyone has the capacity to achieve extraordinary performance in some mode of expression, if each can discover opportunities in a domain of expertise that allows his or her unique set of personal attributes to shine.

But what does it take to identify and develop the raw material of talent and turn it into exceptional accomplishment? How do we parent and educate extraordinarily determined and intelligent children and help them reach their potential? How can we help more conventionally talented children find the self-motivation and external support that moves them toward the fulfillment of their dreams? And how do we shift the course of an educational culture that has, for the past several decades, underchallenged the children it once regarded as its best hope?

Taylor's ostensibly Icarus-like story demonstrates what can happen when we give young people opportunities to rewrite the old myths that have kept our generation (and previous ones) from achieving new heights. It takes considerable courage to overcome our instincts to keep our children's feet on the ground, for we know that if we give them the wings they crave, some will fall. Others, though, will fly; they'll go places and do things—real things—that the mythical Icarus never dreamed of. Some may even discover new ways to soar, as Taylor did, to the sun and even beyond—high enough to capture stars of their own.

PART I

1

THE DIGGER

WHEN I FIRST MEET Taylor Wilson he is sixteen and busy—far too busy, he says, to pursue a driver's license. And so he rides shotgun as his father, Kenneth, zigzags the family's Land Rover up a steep trail in Nevada's Virginia Mountains.

From the back seat, I can see Taylor's gull-like profile, the almost unwavering line from his sandy-blond bangs to his forehead to his prominent nose. His thinness gives him a wraithlike appearance, but when he's lit up about something (as he is most waking moments), he does not seem frail. He has spent the past two hours—the past few days, really—talking, analyzing, breathlessly evangelizing about nukes. We've gone back to the big bang and forward to mutually assured destruction and nuclear winter. In between are fission and fusion, Einstein and Oppenheimer, Chernobyl and Fukushima, matter and antimatter.

Kenneth steers the SUV past a herd of wild mustangs as we climb a series of progressively rougher and narrower dirt roads. This is the third time Taylor has coaxed his dad to these mountains so that he can beef up his collection of uranium ore—part of a broader stockpile of radioactive materials that the teenager has built into one of the most extensive in the world. Kenneth steers around a switchback, flushing a pair of quail, then halts the SUV in front of a small hole dug into the side of a mountain.

"Whoa, wait a minute," Taylor says, throwing open his door.

He leaps out and sprints toward the mine entrance, which is barricaded by a shiny new chainlink fence. "This was my mine!" he shouts. "It was *my* mine, and they fenced it off!"

The Bureau of Mine Safety has hung a sign on the fence: DANGER: UNSAFE MINE — STAY OUT, STAY ALIVE. The smaller print lists some of the dangers in abandoned mines: bad air, rattlesnakes, old explosives, rotten timbers, falling rocks.

"Okay, now, y'all ignore that," Taylor says, calming. He turns toward the truck to fetch the gear, scoffing. "Like any mine is going to be safe."

Taylor "discovered" the Red Bluff Mine the previous year while rifling through a 1953 geology thesis complete with fading Polaroid photos stapled to yellowing paper that he'd found in a forsaken corner of a library at the University of Nevada. Though the mine produced ore commercially for just a few years, the dirt that it cuts through still coughs up, Taylor says, "some of the hottest rocks in Nevada."

Taylor unloads a pickax and a shovel, flashlights, and three types of Geiger counter. He chides his dad for forgetting his radiation-detecting wristwatch and his ore-collecting buckets—"Looks like we'll have to be resourceful," he says—and heads for the fence.

He hoists himself lightly over the top, and Kenneth and I hand the gear to him and then clamber over the chainlink ourselves. When we enter the mine, the Geiger counter's ticking quickens slightly. It's late autumn and unseasonably warm—a good thing, since on warm days uranium mines tend to "exhale" radioactive radon gas generated by uranium's natural decay. In cooler weather, mines "hold their breath," as Taylor puts it, keeping more radon inside.

Taylor fills me in on mine terminology. The Red Bluff opening is an adit, meaning it enters the side of the mountain roughly horizontally (as opposed to a shaft, which enters a mountain at a vertical or steep incline). The darkness pulls in around us as we duck our heads and step inside; I can sense the weight of the mountain above. Swinging our flashlights, we see bats hanging on the support timbers, and rat feces scattered on the ground. (Unmentioned on the sign is the potentially fatal hantavirus, spread via rodent urine and droppings.)

We reach a winze, a side tunnel that angles steeply downward. Though winzes can drop hundreds of feet, Taylor's light follows a sloping plywood chute to another adit only six feet below. He reaches down with his Geiger counter's probe, and the ticking picks up considerably.

"Something interesting down there," he says, already handing his light and radiation detector to his dad. He hops onto the wooden chute and slides down; Kenneth passes the gear to Taylor and we slide down after him.

Taylor quickly finds the radiation source. It's a yellow vein of uranium running diagonally along the brown wall of the tunnel, crossed by a greenish trickle of water. When we move our lights away from the stream, it continues to glow faintly. "Ooh, man, radioactive water,"

Taylor says as he shifts his flashlight beam from side to side, studying the tiny green-gold river from all angles, transfixed. I find myself watching his fascination with a fascination of my own.

"Liquid uranium," the teenager says. "I wonder if it's coming off some autunite up above. It's a fluorescent mineral, hydrated calcium uranyl phosphate; pretty rare 'round here."

We continue deeper into the tunnel until we reach a frail-looking brace. Taylor inspects the rotted wooden beams and cross brace, then shines his light down the curving passageway; the tunnel's end is out of sight.

"We might-could go back farther," Taylor says, using one of the double-modal expressions that attest to his Southern roots. "But it looks unstable to me." Kenneth gratefully concurs, and we retrace our path toward the blast of daylight that meets us at the mine's entrance. Once outside, Taylor climbs the fence and hoists his leg over. As he does, his Geiger counter probe brushes his thigh and emits a loud squawk.

"Huh?" he says. "What's going on with my leg?" He hops down and runs the probe up and down his jeans. The detector shrieks. He looks worried.

"My pant legs are highly radioactive," he says. "This is actually scaring me." He climbs down the other side of the fence and quickly unbuckles his belt. "Uh, Dad, can you run and get the pancake probe real quick?" he says, yanking his belt from its loops and quickly pulling off shoes and jeans. He's standing in his boxer shorts when Kenneth trots back from the Land Rover with the more sensitive instrument. Taylor snatches it from his father's hands and runs the large, flat disk along his bare leg. When it doesn't bleep, Taylor looks relieved. He goes over to the SUV and tests the seats, which are clean. Then he gingerly lifts his jeans and scans them. Halfway down the right thigh, the detector picks up the contamination, an invisible oval patch three or four inches long.

"It's not alpha radiation, which should rule out the mine as a source," Taylor says. "But it also rules out my pants shielding me. I could have absorbed a significant dose. That's kind of embarrassing." He holds the pants up to the sun. "I don't get it. They were clean this morning when I put 'em on. My skin's not radioactive, so it's not loose contamination, which makes me think it's been on the pants for a while. But—how? Generally, my jeans are not radioactive to start the day."

• • •

"Where does it come from?" Kenneth says a few minutes later as we sit in a shady nook watching Taylor dig through the mine's tailings pile. It's a question that Kenneth and Tiffany have asked themselves many times. Kenneth is a Coca-Cola bottler, a skier, an ex–football player. Tiffany is a yoga instructor.

"Neither of us knows a dang thing about science," Kenneth says.

"Sweet Jesus!" Taylor yells from atop the mound of yellow earth. "This is exceedingly radioactive dirt!" He's wearing my spare shorts now, the bunched-up waist cinched, with his belt, around his slender torso. His pickax and shovel lie on the ground next to the clicking Geiger counter as Taylor claws with his hands through the dirt. He bends from his waist, knees locked, his thin, sun-deprived legs descending through swirls of yellow dust and landing inside untied sneakers.

Kenneth squints and wipes a bit of sweat from his forehead as he watches his son dig. He's in his late fifties, tall and solidly built, with smoky-yellow hair that's transitioning to gray. He has a laid-back, aw-shucks sociability acquired at a lifetime's worth of Southern barbecues. Like most of the Arkansas business class, he's conservative in both politics and manners, though he breaks from that mold in terms of open-mindedness. Ask him a question of any consequence and you'll almost always get a short pause as he ponders the matter and then a considered answer.

"Taylor," Kenneth says, "I've got a pair of gloves in the car."

"Don't need 'em," Taylor yells down, his glance carrying a hint of annoyance. "You don't wear gloves when you're prospecting for uranium."

"Why not?" Kenneth asks. I wonder that myself.

"I don't know; it's just not done," he says, continuing to dig. Then, a few seconds later, he adds: "Uranium *is* radioactive, but there's very low radiation activity per the amount of material. You'd have to work in mines a long time for it to hurt you. And it's not soluble, so you can wash it right off; it won't go into solution inside your body."

Taylor takes a couple of whacks with the pick and pulls out a softball-size yellow rock. He checks it with the detector, which screams its approval. "Code yellow!" he shouts joyfully. "Whoa, that's a hottie!"

He sets the rock aside and goes back to digging, using the probe to guide the path of his pick and shovel and, mostly, his hands. The clicks quicken to the point that they become one long bleep. "This is going crazy up here!" he yells, prying out another chunk. "Look at how bright it is!" he says, holding the rock up to the sun as he puts the probe to it. "And appropriately radioactive for its color."

The deeper he goes, the more excited he gets, calling out a play-by-play commentary that veers to flights of fantasy and speculation. "This is the highest-grade uranium I've ever found! I wonder—could this be the infamous natural radian barite, the king of all the hot rocks? Nobody's been able to find it before in the U.S., but who knows? Maybe I'll be the first . . .

"I gotta keep digging!"

Almost from the beginning, it was clear that the older of the Wilsons' two sons would be a difficult child to keep on the ground. "Taylor has always been obsessed with things," Kenneth says as he watches his son scrape away at the earth. "Whatever he got interested in, he just went crazy with it, nonstop. Even getting him to eat was a big trick. Sometimes it still is."

Taylor Ramon Wilson was born in May of 1994 in Texarkana, Arkansas, just north of the Texas-Arkansas border. From the moment he could crawl, he wanted to dig. At their first home, Kenneth built Taylor a sandbox, but it was too constraining for the toddler; he needed a larger swath of diggable terrain. As soon as he could haul himself out of the box, he started tearing up the lawn, digging holes, pouring water in them to make mud, then digging some more.

When he was four, the digging segued into an interest in construction. That in itself isn't unusual for boys. But Taylor the preschooler wanted nothing to do with toy dump trucks or other miniature construction equipment. He played with real traffic cones, real barricades. At age four, he donned a reflective orange vest, yellow boots, and a hard hat, then stood in front of the house directing traffic.

"The neighbors all knew him," Kenneth says. "He'd set up at the side of Wade Trail and stand there with those big gloves waving cars around the barricades. He was shy back then, but that sort of thing seemed to bring him out of his shell. He loved it when people waved or stopped to visit with him."

As Taylor's fifth birthday approached, the family moved to a larger home in a new cul-de-sac neighborhood on Texarkana's far north side. Taylor told his parents he wanted a crane for his birthday. Kenneth brought him to a store and showed him the toy cranes, but Taylor saw that as an act of provocation. "No!" he yelled, stomping his foot. "I want a real one."

This is about the time almost any other father would have put his own foot down. Instead, Kenneth called a friend who owned a construction company, and on Taylor's birthday, a six-ton crane pulled up to the party. The kids sat on the operator's lap and took turns at the controls, guiding the boom as it swung above the rooftops on Northern Hills Drive.

To the assembled parents, all wearing hard hats, the Wilsons' parenting style must have seemed curiously indulgent. Later, when Taylor's interests turned toward more perilous pursuits, the Wilsons' approach to child-rearing would appear to some outsiders as dangerously laissez-faire and even irresponsible.

"Some of what people were saying got back to us," Tiffany says, "and even our friends were sometimes critical, though that usually came through in the form of jokes and kidding. But luckily, other people's opinions don't weigh on us that much. Not if they get in our way of what we want to do."

If Kenneth and Tiffany were winging it at first, as they both admit they were, their parenting strategy was, in fact, evolving into something uncommonly intentional.

What they wanted to do, Kenneth says, was "help our children figure out who they are, and then do everything we could to help them nurture that."

Taylor eventually settled on an interest that would stretch that nurturing capacity to almost inconceivable extremes. But in his preschool years, what Taylor would become was anyone's guess. He hopscotched exuberantly from one infatuation to the next with a deep-focus, serial monogamy. In pictures and videos from that era, Taylor's brother, Joey, three years younger, is typically smiling and engaged, whatever the situation. Taylor, by contrast, looks lost when he's not in costume. But when he's playing a part—excavator operator, archaeologist with metal detector, carpenter with suspenders and tool belt—he looks purposeful and confident.

Outside the Red Bluff Mine, Taylor has chucked his equipment aside and is using his bare hands to scoop out a hole he's been working on for the past half hour. "I think I'm getting closer now to some sort of bulk radiation source," he shouts, continuing his progress report. He scoops and digs some more, exposing the edges of a basketball-size chunk of ore.

"This here's what's gonna make it all worth it today," he says, not looking up. "This could be, like, a thousand-dollar specimen. It may take an archaeological dig to get it out, but we'll manage it—even if we gotta die trying!"

Kenneth chuckles as Taylor grabs the pickax, takes a few whacks, then works the tool under the rock, trying to pry it loose. He watches his son with bemusement, occasionally checking the time and glancing toward the sun, which is settling closer to the western horizon.

"Tay, you fixing to dig all the way to China?"

"If China's got uranium," Taylor says, standing up and looking his dad in the eye, "I will gladly dig that far!"

2

THE PRE-NUCLEAR FAMILY

BOTH TAYLOR AND JOEY were born at home with a midwife, not a common practice in the American South. Midway through her first labor, Tiffany began second-guessing her decision; the baby did not seem to want to come out. "I was panicking," Tiffany says. "But the midwife said, 'Once you start pushing, it will get easy.' "

It didn't get easy. As Tiffany pushed, the midwife reached in and felt Taylor's arm—and promptly got punched. Taylor was pushing back, apparently fighting to stay in the womb. In a struggle that went on for more than an hour, the midwife kept grabbing Taylor's arm, and Taylor kept wriggling away. Finally, she managed to get a grip on him and pull him out.

Moments later, the placenta followed, but it wasn't intact; it was bro ken up, smashed to pieces. "That's when the midwife freaked out," Tiffany says. "She said it was unbelievable and amazing that this baby survived. She kept calling Taylor the miracle baby, and she said she'd never had a kid fight like that. And I do believe," Tiffany says, laughing, "that was one of the last times anyone was able to force Taylor to do anything he didn't want to do."

A slightly built brunette, Tiffany looks about a decade younger than she is. She's fit and energetic and almost unfailingly upbeat, although a small line of worry ("Courtesy of Taylor," she says) has crept between her eyes. Her parents grew up in Hope, Arkansas (the hometown of President Bill Clinton), but moved to Texarkana so Tiffany's father could pursue business opportunities.

Southern Arkansas is deeply conservative country, and Tiffany's father was a strict charismatic Christian. "He'd get out at the picket fence with his Bible and stay all afternoon, preaching sermons," Tiffany says. But her mother, Nell, was a freethinker who reacted to her own conservative upbringing by turning to alternative religions and unconventional approaches to health and wellness. Under her mother's influence, Tiffany grew up as a rare granola child in the 1970s Deep South.

A manic entrepreneur, Robert Bearden spread himself thin with numerous ventures: cattle and construction, nightclubs and barbecue joints. Tiffany's mother believed her husband would work himself to death, but it was Nell who had the first major health scare. When Tiffany was only three, her mother was diagnosed with cervical cancer. The doctors gave her radium implants—and five years to live.

"That changed everything," Tiffany says. "Mom was in her midthirties, and with the cancer, she got really heavy into health-food stuff with me and my sister—although she never could switch my dad. He was a meat-and-potato man and a smoker till the end."

Nell set up Texarkana's first yoga studio, which Tiffany would later expand into a health-food café and juice bar. When Tiffany was in middle school, twelve years after her mother's cancer scare, her father was diagnosed with lung cancer. Within six months, he was dead.

Tiffany is convinced that her parents' lifestyle differences accounted for their contrasting cancer outcomes. She followed her mother more deeply into yoga and health food, and in 1981 she went off to the University of Texas at Arlington, just outside of Dallas, a three-hour drive from Texarkana. "I didn't have any strong interests in any one subject," she says. "To be honest, I just wanted to have fun and stay fairly close to my mom."

Kenneth, too, has deep roots in southwestern Arkansas. His maternal grandparents were farmers; his paternal grandfather brought the family bloodline into the merchant class when he opened a general store in the town of Nashville, Arkansas. After setting up a bottling operation in the store's back room, Forrest Wilson acquired the rights to produce and distribute Coca-Cola in the region. On New Year's Day 1911, a store employee named Hence Wilder filled and capped the first bottle of Coke.

The Coke business took off so fast that Forrest sold the store and threw all his efforts into the bottling operation. His rising fortunes allowed the Wilsons to become—and in some cases literally provide—pillars of the community: their trucks carried the stones that now serve as the foundation for the American Legion Building; their donations built the Nashville Scrappers' baseball park; their scholarships sent local high achievers to college. Forrest cofounded the local Rotary Club and his son, Ramon, cofounded the country club and the annual Peach Blossom Festival. Ramon and his wife, Nelda, were fixtures at nearly every community event, often with Kenneth and their two other children in tow.

Sandy-haired and solidly built, Kenneth was popular, easygoing, and a standout athlete. "It was football, basketball, track and field, and baseball — then back to football," he says. "I never had a break, never wanted one. My life revolved around sports."

When Kenneth was in high school, his grandfather hired him, at fifty cents an hour, to sweep the Coke plant's warehouse and clean its restrooms. The teenager's path was clearly being laid out for him.

At the University of Arkansas, Kenneth joined Kappa Sigma, the campus's oldest and largest fraternity, and majored in business. After graduation, he headed back to Nashville. He settled down, joined the Rotary Club, and in 1988 became the fourth Wilson family member to be company president.

A natural problem solver, Kenneth had already assumed the role of company troubleshooter. "My dad didn't train me; he'd give me a problem and let me solve it. He wouldn't meddle. Some of my mistakes hit his bottom line, but that's how I learned. I got good at finding solutions."

Kenneth's laid-back style seems incongruous with his business acumen and his capacity to build on the advantages he inherited. By 1981 the company's production had grown from seventy-five bottles a day to four hundred bottles a minute. Pursuing more growth, in 1988 Kenneth acquired the regional Dr Pepper soft-drink franchise, and by 1997, sales had quadrupled, and Kenneth's company's territory had the highest per capita Dr Pepper consumption in the world.

On one of Tiffany's frequent weekends home from school, she went out with her cousins to a Texarkana restaurant; Kenneth was there with a group of his friends, and he started a conversation with the petite and cheerful woman nine years his junior. Kenneth had gotten married but was recently divorced, and he told Tiffany he often traveled to Dallas to visit his young daughter, Ashlee, then living there with his ex-wife. Tiffany invited him to give her a call the next time he was in town.

"She was into healthy lifestyles and I was into fried food," Kenneth says, "but before long I started liking the things she liked. And before you knew it I managed to talk her into coming back to live with her mom. Once I got her back to Texarkana, I thought I had a pretty good shot at her."

"On the outside we seemed so different," says Tiffany. "My friends said, 'You're such a health nut and you're going with the Coke man?' But we were a lot alike too. There was just a knowing that we were supposed to be together. And we both wanted to have kids. But that's where we had absolutely no idea what we were getting ourselves into."

It's hard to believe now, but Taylor was at first an intensely self-conscious child, fearful of new situations and nearly paralyzed by shyness. "When he was about to start kindergarten, Kenneth wanted to hold him back," Tiffany says. "He was small and emotionally immature compared to the other kids his age." But Taylor did begin kindergarten that year, with his contemporaries, at St. James Day School.

The Episcopalian elementary school was headed at the time by Dee Miller, who remembers Taylor as timid and unusually empathetic. "That first day of kindergarten, he held a little girl's hand. He stayed with her all day and got her through it, even though he hardly said a word." All through that year and the next year too, Taylor rarely interacted with his classmates except when he was helping them with their work.

It took a national tragedy to pull Taylor out of his shell. In the grieving aftermath of the September 11, 2001, terrorist attacks, the Wilsons traveled to Nashville, Kenneth's hometown, to attend a memorial service. "I think 9/11 really affected him," Kenneth says. "The way the TV kept showing the planes going into the buildings. For a second-grader, it was like it kept happening again and again."

At one point during the event, Taylor, who had up until then sung only at Sunday Methodist church services, suddenly stood up and began singing "God Bless America" in an ethereal soprano. His spontaneous performance transfixed the crowd of several hundred, moving many to tears. Afterward, Arkansas's lieutenant governor, Winthrop "Win" Rockefeller, approached Taylor, removed his own American-flag lapel pin, and silently pinned it to the boy's shirt. Word of the performance spread quickly, and soon people throughout the region were recruiting Taylor to sing at weddings, fundraisers, and other events. He had perfect pitch and enough volume that he rarely needed a microphone.

"He'd put on a little blue tuxedo, and people paid him really well," Tiffany says. "But he would've done it for free. He loved it, and he was so animated, with his hands and his facial expressions."

After September 11, Taylor became intensely patriotic. For Halloween, he dressed as a "rescue hero," carrying a grappling hook as he made his trick-or-treat rounds. He wrote letters of appreciation to workers clearing rubble at Ground Zero, and for his eighth birthday, he requested a cake decorated with a flag and the words *God Bless America.* Taylor joined the Cub Scouts and later the Boy Scouts, happily wearing the uniform, saluting the American flag, and going through the Scouts' quasi-military drills.

Just before his ninth birthday, Taylor dropped his construction tools and announced that he was going to be an astronaut. Tiffany baked him a rocket-shaped organic cake and Taylor started reading everything he could get his hands on about space exploration. He also began sending letters to astronauts, requesting pictures and autographs. At bedtime now, he would cast aside any book that ventured into the realm of fantasy. He no longer wanted to know about heroes and magic; he wanted to know how manmade things traveled through air and space.

Tiffany and Kenneth did their best to keep up with their son's rat-a-tat-tat questions but quickly reached the point where they lacked answers to his increasingly complex queries. They brought him to the Texarkana Public Library, where Taylor asked the reference librarian what she had on rockets and space. "If she'd been honest she would've told me not to waste my time there," Taylor says. "It was a pretty disappointing place if you wanted to find out anything about anything. Even for a third-grader."

The local hobby shop proved more enlightening. Kenneth and Taylor began building small rockets and launching them on weekends, when the sprawling parking lots at Four States Fairgrounds were usually empty. "We'd bring Joey along," Taylor remembers, "and he just loved it. When the parachute came out, me and him would run and try to catch the rocket when it floated down."

That summer, Taylor spent his weekdays building and repairing his rockets. By midweek, he'd grow restless; his weekend launch window would seem a hundred years away. Taylor began pestering his maternal grandmother, who had moved into the house next door, to take him out on weekdays to launch rockets.

"Taylor and my mom were extremely close," Tiffany says. "She kept an eye on him and she knew him like a book; maybe better than anyone else."

To demonstrate that he knew what he was doing, Taylor set up his launch apparatus in the driveway and took his grandmother through all the prelaunch steps and safety checks. Finally convinced, Nell drove her grandson and his rockets out to the empty fairgrounds on a windless Wednesday afternoon in August.

The first rocket up for launching was one Taylor had repaired after a hard landing. With its broken fin glued back on, it looked as good as new, and Taylor inserted the igniter into the bottom of the engine and slid the fuselage down the launch rod and onto the pad. Then he backed away to the controller, where his grandmother was waiting.

Taylor armed the firing switch and counted down dramatically—"Five, four, three, two, one"—then pushed the launch button. "Blastoff!" he yelled as the rocket hissed off the pad—leaving its no-longer-glued-on fin behind. Once the rocket cleared the rod, it veered sideways and whizzed horizontally toward Taylor and his grandmother.

"It was like it was aiming at us!" Taylor remembers.

There was no time to hit the ground, no time to move. The rocket whooshed between Taylor and his grandmother, standing just five feet apart. It skittered across the field and tunneled into the grass, then jettisoned its nose cone and parachute with a final pop.

"At that point," Taylor remembers, "we decided to pack up the rockets and call it a day. On our way home, Grandma and I agreed that it was probably best to keep the whole thing quiet."

3

PROPULSION!

"THERE ARE FIVE THINGS involved in a space shuttle launch," Taylor begins, spinning around to face his third-grade classmates, his left hand behind his back, his right index finger raised emphatically. Taylor, wearing a blue astronaut jumpsuit with NASA patches, paces in front of a triptych tabletop poster titled "Aeronautical Engineer/Astronaut" and then comes to an abrupt halt. He picks up a plastic model of the space shuttle.

"First, sparks are sprayed to make sure there's no hydrogen in the air. When everything's lighted, it sways back and forth and then it takes off." Taylor sweeps the model through the air. "Then Kennedy Space Center turns it over to Johnson, and Johnson calls for the roll maneuver."

"Johnson?" a girl's voice asks. The video camera pans left to a curly-haired girl with her hand raised.

"Houston," Taylor says.

"Like as in 'Houston, we have a problem'?" the girl asks.

"Yes," Taylor says tersely. "A couple seconds after it breaks through the sound barrier, the boosters fall off." He tries to snap apart the model but it doesn't separate, so he just points to the components. "After it breaks through the ozone [layer], the external fuel tank falls off. Then, if you're gonna orbit the Earth, okay, you gotta turn," he says, acting it out. "But if you're going to the ISS, you gotta make a much larger turn."

"What's the ISS?" two kids ask.

Taylor jumps up into the air. "The International Space Station!" he shouts. He replants both feet on the floor—briefly—and adds, "For some reason, NASA uses everything in abbreviations. Speaking of abbreviations, who knows what NASA stands for?" Hands shoot up, and Taylor points to one acronym-guessing kid after another, clearly enjoying shooting down their answers. Finally, he blurts out, "It's the National Air and Space Administration!"

Though the budding astronaut has botched the second word (it's

Aeronautics), he's already moving on, pointing to pictures taped to the board. "This is me at Kennedy Space Center. This is me and my dad in the assembly building. This is me with an astronaut. Here are all the *Columbia* astronauts who died. And here," he says, solemnly picking up an object wrapped in clear plastic, "is a ceramic tile made for *Columbia,* which I got at Cape Canaveral. It's highly toxic, so don't open it."

"Highly toxic?" one kid says. "Open it!"

"Would you die?" another kid asks.

"No," Taylor says. "You'd just burn."

His classmates rapid-fire questions at him.

"Have you ever gone up into space?"

"Have you ever tasted astronaut ice cream?"

"What about the astronauts who died?"

"Do you know that your dad is taping all this on video?"

"Shush!" Taylor yells. "I can't take all these questions!" He turns around, grabbing at his head, then spins toward the class. "I'm not answering any more till everyone gets quiet!" He sits on the table, smiling and shaking his head as the kids laugh away. In half a minute, they're quiet. Taylor stands up and points to Ellen Orr, the curly-haired girl in the front row, who has been frantically waving her hand.

"Taylor, why do you want to *do* this?" she asks.

"You can help people," Taylor says, "and make breakthroughs in outer space. And if you're the engineer, you can build safer planes and shuttles. And if you die in a tragedy, you'll still be a hero."

"The thing about Taylor that no one's ever seen with any other student," says St. James's former head of school Dee Miller, "is his *passion.* He was obsessed with science and he poured himself into everything he did. Everyone loved watching him, and when he was in the room, he just took control."

"This will sound strange," says sixth-grade teacher Angela Melde, "but I'd say Taylor was the only true Renaissance man I've ever met —and he was a grade-schooler."

The carbon composite tile Taylor passed around wasn't actually toxic. "That was complete BS," the teenage Taylor would confide. "I just didn't want kids opening it up and touching my tile." But it *was* one of thousands of tiles made for *Columbia,* the shuttle whose tile-coated wing was breached by a falling chunk of foam insulation, compromising its ability to hold up to the heat of reentry.

Just two months before Taylor's presentation, *Columbia* had disintegrated as it returned to Earth, scattering debris across Texas, Arkansas, and Louisiana. For the next few weeks, Taylor had one parent or the other drive him out after school to search for pieces of the wreckage. Authorities had warned the public not to touch anything because the parts could be radioactive, so Taylor brought along a toy robot grabber-arm from the NASA gift shop, which he figured he'd use if he found something. "So I wouldn't actually touch it, but I *would* pick it up."

Taylor had studied aerospace accidents from a scientific angle, but *Columbia* was, he says, "an emotional thing; it had a big impact on me. Space was my *thing;* I actually felt like I was part of the NASA family. I wasn't scared or deterred by the accident, but it made me think about the responsibility that people who build things like the shuttle have. More than anything, I felt like I needed to be involved to make sure it didn't happen again."

As the end of Taylor's third-grade year approached, *astronaut* was still the first word that came out when anyone asked him what he wanted to be when he grew up. But it was becoming increasingly clear that Taylor didn't want to just *ride* in space chariots; he wanted to build them too. Now he was tweaking his kit rockets, playing with different engines and wrappings, fine-tuning fins, experimenting with ways to minimize drag, maximize thrust, boost altitude. Before long he'd moved beyond kits and was building rockets from scratch. With his grandmother's blessing, he took over a corner of her garage and set up a table that was soon strewn with tubes, parachutes, and engines. Each afternoon after school, he'd jump out of his mom's car and sprint toward his rocket laboratory, Tiffany's shouted reminders to have a snack fading behind him.

Taylor had been learning mostly by trial and error, but the pace of his progress was frustrating. If his rockets were going to fly higher, he'd need to more precisely tune the aerodynamics and weight dynamics. Among the trickier challenges was keeping the center of gravity ahead of the center of pressure (the center point of the aerodynamic forces), which can be complicated, since both centers change constantly throughout a flight. To inform his technique, he sent away for advanced books and consulted rocketry sites on the Internet. He learned how to calculate center of pressure and find the velocity a given rocket needs to be aerodynamically stable as it comes off the launch rod. Once he knew how much acceleration was necessary, he could figure out the required thrust and then pair the rocket with the right motor.

At the time, Taylor says, he didn't think of what he was doing as applied physics. "I just wanted my rockets to go higher."

But as his hand-built rockets got bigger, Taylor started running up against the cost and thrust limitations of hobby-shop engines. "What I really wanted to do," he says, "was build me those big rockets they shoot off in the Black Rock Desert, where they need to get FAA [Federal Aviation Administration] clearance to launch." Hoping to stretch his allowance and singing money, Taylor began to reverse-engineer his store-bought engines, taking them apart and analyzing their nozzles, propellants, and ignition systems. "At that point," he says, "I realized I could probably make my own engines, even my own fuels."

To do that, he'd need to delve deeper into the raw materials of propulsion. He'd need to break down propellants into their components and understand, on a molecular level, how they combined and interacted. In other words, he'd need to master chemistry. Taylor pinned up two posters of the periodic table of the elements, one in his bedroom and one in his garage rocket factory. Within a week he'd memorized all the elements' atomic numbers, masses, and melting points. That summer, Kenneth's company was consolidating production with another distributor and moving the bottling line offsite. Whenever Taylor visited the Coke plant, he'd notice a piece of surplus equipment he could use for his lab and would pester his dad to let him bring it home.

Taylor brought a second table into the garage and topped it with his newly acquired burners, flasks, incubators, and petri dishes. When his parents upgraded their kitchen, Taylor claimed the old stove. What he couldn't find at the Coke plant, his grandmother would usually buy him, especially safety gear and books. By the time school started again, Taylor's lab had taken over the entire garage, and his grandmother's car was relegated to the driveway. Nell told Tiffany and Kenneth that she didn't mind; she was happy to encourage her grandson's hobby. That Christmas, she bought Taylor a white chemist's lab coat with his name embroidered over the pocket.

One of the few discoveries Taylor made at the Texarkana Public Library was a reference to an article in *Scientific American* describing a way to image an ant's brain using a microscope and household chemicals.

Taylor learned that the article was part of the magazine's classic Amateur Scientist columns. For more than seventy years, the series had shown do-it-yourselfers how to create sophisticated science projects on an amateur's budget. Credited (along with the space race and various governmental and philanthropic initiatives) with fueling the mid-twentieth-century citizen-science movement, C. L. Stong's and Jearl Walker's DIY articles on how to build lasers, electron microscopes, and even a cyclotron kick-started many a science-fair project and inspired countless young people to pursue scientific careers.

"When I got ahold of a secondhand CD-ROM collection of those columns, my life changed," Taylor says. "I realized there were these world-class experiments that cost millions in top laboratories that you could replicate at home." Taylor's lab work shifted into high gear. One week, he was culturing cell lines; the next, he was sorting molecules with electricity or producing glow-in-the-dark bacteria by isolating the gene that made jellyfish bioluminescent and inserting it into *E. coli* cells. When he needed a chemical he couldn't buy—such as crystalline iodine, a powerful reducing agent useful in many chemistry applications, including illegal methamphetamine production—he figured out how to synthesize his own by combining and processing other chemicals.

Following the magazine column's recipe, Taylor started his own biodiesel-production facility. He tried to convince Kenneth to let him run the Coca-Cola fleet off his homebrewed fuel, "but my dad made a cost justification that if I ruined a vehicle, it would be thousands of dollars."

Mostly, though, Taylor was experimenting with propellants for his rockets, one of which was his biggest yet, a four-foot-long monster nearly as tall as the nine-year-old himself. "It was sort of like *Rocket Boys,* but a one-man show," Taylor says, referring to Homer Hickam's memoir (adapted into the movie *October Sky*) about teenage rocketeers in a 1950s Tennessee mining town.

Like Hickam and his buddies, Taylor tried dozens of combinations of fuels and oxidizers: various black powder mixes; ammonium perchlorate composites such as those used on the space shuttle; even "rocket candy," a blend of powdered sugar and an oxidizer such as potassium nitrate. "It's basically edible rocket fuel," Taylor explains, "and it's actually quite energetic—and a whole lot cheaper than anything you can buy off the shelf."

For rocket candy to achieve the proper, putty-like consistency, its ingredients must be heated, dissolved, and mixed correctly. Taylor experimented with various fuel/oxidizer ratios, trying to pinpoint the perfectly peppy mix, but his rocket-candy propulsion program sputtered when a boil-over ruined his lab's stove. A few weeks later, he wrecked a secondhand replacement stove his grandmother had bought him.

Taylor was getting impatient. One afternoon, he decided to make some rocket fuel, stove or no stove. Inside his lab, he filled a Coke can with his carefully measured ingredients. Then he put on his lab coat and his polycarbonate face shield and went outside with a propane torch and some tongs. "Part of me understood that it wasn't the best thing to do with a torch," he recounts. "But I'd already wrecked the stove, so I was thinking, *What else can I do?* At least I was outside and wearing a face shield."

Rocket candy is, of course, extremely flammable, and just about every guide to making it warns you not to use an open flame or any direct heat, since uneven heating can auto-ignite the volatile ingredients.

"I was running the torch along the bottom of the can," Taylor says, "and at the same time trying to mix it with a spatula, and—well, you know where the story is going . . ."

When the fireball cleared, Taylor looked down and saw that the can was gone—vaporized. The explosion burned his hands slightly (he still has small scars) and took the hair off his arms.

"But the worst thing," he says, "was that I had spectators."

The "spectators" were his parents. Taylor had convinced them that what he was doing was safe, and they'd come to watch. True, he had worn his facemask and backed them away. But he obviously wasn't doing everything he needed to protect himself. As responsible parents, Tiffany and Kenneth agreed that night, they needed to put constraints on their son's experiments with fire and volatile chemicals.

The next day, determined to maintain a strict and united front, they sat Taylor down. "But when we went to talk to him," Kenneth says, "Taylor broke the whole thing down like he was on the NTSB [National Transportation Safety Board] analyzing some major accident." In a point-by-point presentation, Taylor went over every aspect of his technique that had fallen short in terms of safety and then showed them the systematic plan he would implement to improve his methods and make them safer.

"We went in dead-set on reeling him in," Tiffany says.

"But," Kenneth says, "the truth is that we've never been able to."

4

SPACE CAMP

IT'S A SEVEN-HOUR DRIVE from Texarkana to Huntsville, Alabama, the birthplace of America's space program. As Kenneth piloted the car, Taylor talked nonstop about the things they'd see and do at the U.S. Space and Rocket Center, where they had signed up for a three-day program of hands-on science activities called Space Camp.

To open up learning opportunities for Taylor beyond the immediate area, Kenneth had begun taking his son on educational road trips. They first went east to the U.S. Astronaut Hall of Fame, in Florida, where Taylor talked with Richard Gordon Jr., who had walked in space on Gemini 11 and orbited the moon on Apollo 12. They traveled south to Houston to tour the Johnson Space Center, NASA's center for training, research, and mission command and control. Then the entire family went to Cape Canaveral to watch a space shuttle launch. Unfortunately, the liftoff was scrubbed during the countdown's last minute.

Kenneth and Taylor were still several miles from Huntsville when they caught their first glimpse of the U.S. Space and Rocket Center's imposing centerpiece, an upright, full-scale replica of a Saturn V rocket. They arrived just before noon and spent the next few hours exploring. Everywhere Taylor turned, there was something to discover. There was Wernher von Braun's notorious V-2 rocket, the first manmade object ever to go into space, as well as his Mercury-Redstone and Jupiter-C rockets, which put the first American satellites and men into space. With his father trotting behind, Taylor darted from an Atlas rocket lying on its side as if resting up for another launch to the original Mercury and Gemini capsule trainers to a lunar rover. They peeked into rocket engines of all eras and sizes, and gazed up at the space shuttle *Pathfinder,* complete with its booster rockets and fuel tank. They ducked their heads inside Apollo 12's Mobile Quarantine Facility, where returning astronauts stayed until doctors were convinced that they hadn't come home with any moon bugs.

Just about everyone who was cognizant on April 12, 1961, remembers the day human space flight became a reality. Yuri Gagarin's 108-minute trip around the Earth was an intensely symbolic benchmark in the space race, which had begun three and a half years earlier, in October of 1957, with the Soviet Union's launch of Sputnik 1, the first artificial Earth satellite. The American military had answered two months later with a heavily promoted live TV broadcast of the first U.S. attempt to shoot a satellite into orbit. As millions of Americans watched, the Vanguard's booster ignited and the rocket rose a few feet off its pad, then it lost thrust, exploded, and toppled over into a fireball.

In the heart of the duck-and-cover days of the Cold War, these events filled Americans with dread and confusion. How could the Soviet Union have come off the starting block so far ahead of them? Was it possible that the ideologically inferior Soviets were actually technologically superior? And if the Russians managed to seize control of space, what might they take control of next? Commentators speculated about the impending end of the American way of life and, perhaps, Western civilization itself.

Catching up was a matter of both national security and national pride. The U.S. military commanded ever-increasing resources for aircraft, spacecraft, and a new class of weapons—ballistic missiles—invented by Wernher von Braun, the enigmatic German engineer who had been spirited out of his country by American intelligence operatives, who also stripped his biography of its pesky moral shortcomings. Credited as "the Father of Rocket Science," von Braun left his Nazi past behind and eventually became technical director of NASA's Marshall Space Flight Center in Huntsville, overseeing the town's rapid transition from the watercress capital of the world to the sprawling, high-tech Rocket City.

After the Vanguard disaster in December 1957, public hysteria and political pressure compelled the mortified military leadership to give von Braun the go-ahead to build the orbiter rocket he'd proposed years earlier. On January 31, 1958, within three months of the Sputnik launch, von Braun's rocket blasted the Explorer 1 satellite into orbit. But the rising German American star had bigger things in mind. Shortly before the Soviets orbited a payload of 2,925 pounds in May of 1958, von Braun proposed building a superbooster, called the Saturn, which would be able to lift large loads of men and machinery into and beyond the Earth's orbit. In 1960, the military leadership asked von Braun to take charge of NASA's propulsion programs, and they gave his Saturn project the go-ahead.

Things were abuzz on the ground too, as educators scrambled to bring American children's math and science skills up to par with the presumably better-educated and better-disciplined Russian kids'. Worried teachers pushed their students to study harder, and Sputnik-inspired homework assignments began to pile up. School systems around the country introduced more challenging curricula, pumped up professional-development programs for science teachers, and poured money into upgrading long-neglected science labs.

For the first time, nurturing the talents of the brightest students, who were suddenly seen as a strategic resource, became a nationwide priority. Educators realized that gifted children had special needs and argued that meeting those needs was America's best long-term bet to catch up with the Soviets. In the 1960s, accelerated learning programs and whole schools for academically talented children sprang up all over the country. These programs quickly began to pay off as Sputnik-generation brainiacs rose from primary and secondary schools to college undergraduate programs and beyond. A surge of highly qualified science and engineering students filled university graduate schools, fueling an unprecedented increase in PhDs in the 1960s that wouldn't peak until the early 1970s.

Apart from its intensity, Taylor's obsession with space wasn't all that unusual. For more than forty years, American children named *astronaut* as one of their top-ten dream jobs, according to the Marist Poll. Space continues to beckon adventurous young minds in a way that more grounded pursuits (and they are all, by definition, more grounded) can't match. For boys, astronaut has been the number-one career choice for decades; for girls, it became popular after 1983, when Sally Ride became the first American woman to fly in space. The popularity of the profession began to fall in the U.S. only in 2011, when NASA announced that its last space shuttle launch was approaching. Paradoxically, *astronaut* continued to be named as a top-ten career choice among children in the United Kingdom, even though the British space program has always focused exclusively on unmanned missions.

I wanted to be an astronaut too, of course. But when it came to bridging the gap between imagination and action, I had a radically different experience than Taylor. The idea of tossing time and effort into supporting a child's hands-on pursuit of a highly improbable dream was something that wouldn't have occurred to my mother or father, or to most parents in their circle. Even if my parents had had the resources, it's hard to imagine them hauling me to Florida to meet an astronaut or to Alabama to bounce in a zero-gravity simulator. Nor would it have occurred to them to help me connect the dots between my aspirations and Newtonian physics or to locate mentors who could point out practical paths toward my goals. As for building and launching my own rockets . . . well, that would have been an amusing conversation. Real rockets, the ones that actually go up, were too expensive, too hard to build, too dangerous. Instead, my parents bought me an Apollo poster and a plastic Saturn V model, the kind you glue together and set atop a shelf. I found a scrapbook and pasted into it every space-related newspaper or magazine clipping I could find, keeping meticulous track of the race to the moon. That scrapbook was my pride and joy, but the unspoken message contained within it, and within my models and posters, was that space travel was a spectator sport — and that I was destined to be a bystander, not a participant.

That, of course, is an entirely reasonable conclusion if the hard numbers of probability are one's only guide. As of early 2015, only 541 people have been to space (which the Fédération Aéronautique Internationale defines as beginning one hundred kilometers — sixty-two miles — above sea level). Astronauts are a rarefied class. Even for those aspirants who get through the daunting prerequisites — engineering, physical sciences, mathematics, and (for those on the flight deck) at least a thousand hours of pilot-in-command time in jet aircraft — the odds are long. The notion that Taylor or any kid could eventually make a career out of exploring the cosmos firsthand is not very realistic. But realism isn't what young children need. They need experiences that unfurl their dreams, especially those big, sparkling dreams that inspire them to explore and learn and discover and grow — and that often open up real-life possibilities.

Wernher von Braun, for all the heavy ethical baggage he carried, seems to have understood this. During the final months that he was refining the Saturn V, his crowning technical achievement, the Father of Rocket Science was preparing to launch an ambitious educational project. Von Braun proposed building a museum complex that would go beyond showcasing his propulsion triumphs and preserving the hardware of space exploration. Years before the term *interactive* became a museum buzzword, von Braun envisioned a hands-on space camp to train children in science. As he imagined it, the camp's active programs would encourage intellectual development in the same way that youth sports camps encouraged physical development. He reasoned that giving children an immersive introduction to the excitement of space exploration would inspire them to pursue careers in aerospace or other specialties of science, technology, engineering, and mathematics — the so-called STEM subjects that American kids had studied with such zeal during the immediate post-Sputnik era but had since fallen behind in.

Thus far, only two of the six hundred thousand Space Camp graduates have gone on to become actual astronauts. "But that's not what it's really all about," says Space Camp's former education-curriculum specialist Julie Hatton. "It's about getting kids fired up about science. Children come in to be an astronaut for a day or a weekend, and they discover that they really like science. Then they go home motivated to do all sorts of things, whether it's aerospace or engineering, or some other scientific or technical discipline."

The final stop on Kenneth and Taylor's orientation tour was the massive building that houses the real, restored Saturn V, suspended horizontally above the floor. "You could actually smell it when you walked through the door," Kenneth says, "even though they'd emptied the tanks. It smelled big and formidable, and it looked even bigger."

The father and son and their group of Space Campers followed their guide as she led them along the rocket, starting at what had been its top. The group peered up at the command module and the encased lunar module, then continued under the third stage and the longer second stage. They walked along the massive first stage whose tanks had held hundreds of thousands of gallons of kerosene and liquid oxygen.

When they arrived at the base of the first stage with its five F-1 engines, Taylor fell uncharacteristically silent. The guide was describing the engines' specifications, but Taylor tuned her out, privately studying the huge exhaust nozzles. He cocked his head and peered inside, his imagination drawing him past the outer nozzle and into the turbine exhaust manifold, then deeper into the thrust chamber, where the oxygen and kerosene united and ignited, generating enough firepower to push the rocket's fourteen hundred tons up and away from the Earth.

It was, Taylor would say later, "the most exciting moment of my life."

In years to come, I'd hear him describe many moments of his life that way. But here, for the first time, the boy found himself in the place he'd long dreamed of: peering into the heart of propulsion incarnate. For the first waking moment since his son could talk, as far as Kenneth could remember, Taylor was speechless. He stayed that way for several minutes, studying the engines in silence. Then he turned and raised his hand.

Minutes later, as the docent ran off to fetch her boss, one of the parents asked Taylor why the second and third stages burned liquid hydrogen rather than the kerosene that fueled the first stage. Taylor explained that kerosene was capable of more short-term thrust, so it could get the rocket quickly beyond the dense air and high-gravity zone. But hydrogen has a higher specific impulse (change in momentum per mass), so von Braun chose it to propel the spaceship through the less-resistant higher altitudes. Taylor was answering another parent's question regarding the comparative advantages of the Russians' ballistic missiles versus the Americans' when the director arrived, freeing the rest of the group to move ahead with the guide.

"The director realized Taylor knew as much about rocket history as he did about the physics," says Kenneth, "so he started calling in other people from around the museum, history folks and experts on this and that." Kenneth watched and listened as his son held court with the adults. For the next forty-five minutes, they all went back and forth about the payloads of various rockets, about the European space program, about the progress of the Ares rocket (then under development in Huntsville but since canceled). They also talked about the *Columbia* disaster, which was still fresh in everyone's mind.

Taylor and Kenneth would have a great time over the next two and a half days working their way through Space Camp's ultra-immersive activities. They'd strap themselves into a device that simulated the gravitational force on the moon and make giant bunny hops across the floor. They'd launch small rockets and join in group experiments. Taylor would take on the role of mission specialist during the final exercise, an orbital mission in the space shuttle's cockpit simulator.

After the Sunday afternoon graduation, father and son drove back to Texarkana, where Taylor told his mother and brother and grandmother that he'd had the most exciting weekend of his life. What's more, he'd come home with a headful of ideas — for his rockets, his laboratory, and his future.

5

THE "RESPONSIBLE" RADIOACTIVE BOY SCOUT

As its name makes plain, Texarkana spans the Texas-Arkansas border, just a few miles from the northwest corner of Louisiana. Texarkana has two of just about everything: two mayors, two police and fire departments, even two slogans—"Texarkana is twice as nice" and "Where life is so large it takes two states."

Set at the junction of four major rail lines, Texarkana reached its apogee in the 1950s. Tiffany can point out the famous Hut Club (now called the Arkansas Municipal Auditorium), which her father owned in the mid-1950s when a young Elvis Presley stopped in for an audition. "My dad apparently was not impressed," Tiffany says. "After a couple of songs, he threw him out. When Mom found out Dad treated Elvis that way she scolded him something furious. I think my dad might've been a little jealous of Elvis."

Once grand but now empty hotels haunt the downtown skyline, and along mostly empty sidewalks, young men wander aimlessly past boarded-up shop fronts. There are a few enticing barbecue joints, but other intermittent signs of commercial life are more dubious: Boll Weevil Pawn Used Cars, Edge-Texas Concealed Handgun. A first-time visitor can't escape the impression that most of this twin-town's good days are behind it.

As you head north, out of downtown, the abandoned buildings give way to sterile strip malls backed by cul-de-sac neighborhoods that define Southern rural suburban sprawl. The Wilsons' house, on a double lot with seven acres of lawn and groomed forest, seems far from downtown. You can hear a nearby interstate freeway, just barely, but in the intense heat of midsummer, what you mostly hear are the hum of central air conditioners and the chatter of sprinklers cooling large lawns.

During the summer between Taylor's third and fourth grades, new sounds began to bounce around the neighborhood. One evening, Taylor invited Tiffany, Kenneth, and Joey out to the backyard, where he dramatically held up a pill bottle packed with a mixture of stump remover (potassium nitrate), which he'd discovered in the garage, and sugar. He set the bottle down and, with a showman's flourish, lit the fuse. What happened next—and it happened so quickly that no one had time to think—was not the firecracker's bang everyone expected but a thunderous blast that brought panicked neighbors running from their houses. Looking up, they watched as a small mushroom cloud rose, unsettlingly, over the Wilsons' yard.

After Taylor's stump-remover bomb, big bangs became an almost-nightly part of life on Northern Hills Drive. "You could divide the neighbors into two camps," Taylor says. "The ones who freaked out and pulled their kids inside and closed the blinds, and the ones who came over to watch. But the surprising thing was, no one ever called the police."

Ever since the Chinese started messing around with gunpowder-filled bamboo tubes more than a thousand years ago, the line between fireworks and rockets has been a thin one. For his part, Taylor loved rockets, and he loved explosions; it was only a matter of time before he would combine them.

As Taylor's command of pyrotechnics progressed, he began putting on increasingly extensive fireworks shows. Starting with store-bought rockets, he progressed to making his own, mixing batches of propellants in five-gallon buckets and packing the fuels into his homemade rockets and aerial shells. He figured out which elements would produce the best colors when they went up and exploded: strontium for red, arsenic for yellow, lithium for red and blue, copper and barium for green. Lying in bed at night, Taylor would dream up new combinations to experiment with the next day. "I'd throw in a little extra of this or that to slow the burn rate and get those long sparkling streamers or colored smoke trails."

Taylor spent the entire two weeks before one Independence Day working on his show. "There was a lot of anticipation that night," Kenneth says. "We had five families over, and Taylor had a big pile of stuff ready."

Taylor plucked a rocket out of the pile and brought it over to the plywood launch pad. With a fire extinguisher ready and his face shield flipped down, he flicked his lighter, lit the fuse, and backed away.

"There must have been an air pocket inside the fuel," Taylor says. The rocket came off the pad, sputtered and tipped over, then reignited.

It shot sideways, first toward the woods, then straight at Taylor's stack of fireworks.

"It landed right in the middle of 'em," Taylor says. "We all knew exactly what was about to happen."

"You've never seen such a scattering," Kenneth remembers. "Things flying in every direction, people running everywhere, getting chased by rockets. We were all diving behind anything we could find."

"It was sheer luck no one got hurt," Taylor says. "It makes me think of something they said when we visited Cape Canaveral, about how the NASA engineers learn more from their failures than their successes."

Unlike NASA's, though, the goals of Taylor's rocketry program morphed from seeing how high he could fly his creations to seeing how artfully he could destroy them. The quest for altitude became passé—his rockets needed only to clear the treetops so the entire neighborhood could see them explode into multicolored bursts. His Fourth of July shows got more popular and more sophisticated; with electronic ignition, Taylor could set up synchronized chain reactions that lasted up to twenty minutes.

"After the grand finale," says Taylor, "you'd hear all the neighbors out all around, clapping and hooting."

It was the same sort of adulation he'd gotten from singing, which continued to fund his science hobbies. Taylor loved being the center of attention; he loved the applause and the awed and admiring faces. With his voice changing from a delicate soprano to a harder-to-control tenor, Taylor began to realize that he could use chemistry—especially the more volatile reactions—to catalyze the same kind of responses he'd gotten from his vocal performances.

It's tempting to interpret Taylor's grade-school transformation from a timid, shrinking kindergartner to a charismatic showman of song and science as a rather extreme overcompensation for deep-rooted insecurities. And some psychologists would argue that Taylor the teenager is *still* trying to meet the needs of his much younger self by seeking adulation through his high-profile overachievements.

"I don't think that's it," says Taylor's childhood friend Ellen Orr. "No one who went to St. James school had to do that. Taylor was always accepted and no one ever picked on him; we loved his eccentricities. There were fifteen of us who grew up together and we were *all* very social and expressive. We helped each other learn, we grew up in an environment that fostered confidence."

Taylor, for his part, doesn't remember ever being shy. "I think all the singing helped make me a good public speaker," he says. "But honestly, I can't think of a moment of my life when I've been shy or afraid of what people would think," he would tell me in 2012 just before giving his first TED talk. "Confidence has just never been an issue. I love speaking and performing and I love telling people what I'm doing."

Taylor also loved the process of figuring things out. This is a trait he shares with his brother, who showed an early flair for math. But in terms of revealing their discoveries to the world, Taylor and Joey were on opposite ends of the spectrum. As Taylor increasingly sought the spotlight, Joey became more introverted, preferring to spend time by himself or with a couple of close friends.

"Joey doesn't want everyone to know what he's doing," says Tiffany. "He likes to keep his talents to himself." When his teachers encouraged him to enter mathematics competitions and science fairs, Joey wasn't interested. Tiffany says on several occasions she has come into the bathroom after Joey's been in the shower and found the glass door filled with calculus equations drawn in the steam.

"I think Taylor sometimes embarrasses Joey," Tiffany tells me.

"I wouldn't say he embarrasses me," Joey says. "But growing up with him was hard. He can be loud, always yelling or playing music. I like it better when things are quieter. There's so much drama when Taylor's around. He makes things happen, but he takes up a lot of space and energy."

For Taylor's tenth birthday, his grandmother Nell took him to a nearby Books-A-Million store and invited him to pick out anything he wanted. Taylor made a beeline for the science section, of course. The book that caught his eye was *The Radioactive Boy Scout*, by Ken Silverstein. "My mom bought it for him," Tiffany says. "And she regretted it until the day she died."

The book tells the disquieting tale of David Hahn, a teenager in suburban Detroit who in the mid-1990s attempted to build a nuclear breeder reactor in a backyard shed, with nearly disastrous results.

Taylor was so excited by the story that he read much of it aloud to his family. They heard about the seventeen-year-old Hahn raiding smoke detectors for radioactive americium, the neutron gun he built to irradiate materials, the breeder reactor he cobbled together in an attempt to generate fissionable material, the eerie blue glow emanating from Hahn's makeshift laboratory late at night, the terrified neighbors watching as a Superfund cleanup team in hazmat suits hauled away the family's contaminated belongings. Kenneth and Tiffany saw Hahn's story as a cautionary tale. But Taylor, who had recently taken a particular interest in the heavy, unstable elements near the bottom of the periodic table, read it as a challenge.

"Know what?" he told his parents. "The things that kid was trying to do, I'm pretty sure I could actually do them."

Those were not comforting words for Kenneth and Tiffany. "I grew up during the Cold War," Kenneth says. "I lived through the Bay of Pigs and the embargo, and I was scared of nuclear stuff." Tiffany's knowledge of nukes was limited to newspaper and TV stories about Chernobyl and Three Mile Island—and, closer to home, the so-called Damascus Accident, among the most harrowing near-misses in the history of U.S. nuclear weapons. In 1980, at a ballistic missile launch complex in central Arkansas, a socket dropped from a wrench punched a hole in a Titan II missile's fuel tank. Nine hours later, the complex exploded, ejecting what was then the most powerful thermonuclear weapon in the U.S. arsenal up and through the 740-ton launch door. The W-53 warhead, built to create a blast six hundred times more powerful than the one that leveled Hiroshima, landed in a nearby field. Its safety mechanisms worked as designed, failing to detect the intentional sequence of events that would have armed the bomb and enabled a nuclear detonation.

David Hahn, like Taylor, had gotten into chemistry in grade school, in his case after his step-grandfather gave him *The Golden Book of Chemistry Experiments*. David's father, Ken, was an engineer at General Motors, as was his second wife, Kathy Missig. Ken Hahn, an emotionally distant man, was at first pleased when he found his college chemistry textbooks relocated to his twelve-year-old son's bedroom, where David had set up a small laboratory.

David's parents had divorced when he was a toddler, and the chaotic aftermath of remarriages and shuttling between households had affected David deeply. Shy, gangly, and unpopular as he entered his teen years, he immersed himself in chemistry. "It was an escape," he told me, "and even though the science teachers at school liked me, the experiments at school were too simple; I could do cooler stuff at home."

Like Taylor, Hahn became interested in the more volatile chemical combinations; by his freshman year in high school, he was making his own nitroglycerin. But after several chemical spills and explosions at the house, Kathy and Ken forced David to move his laboratory to the basement and forbade all pyrotechnic experiments. Kathy began making regular raids on his bedroom.

"She was a little crazy and obsessed," David said. "I'd hide sodium metal and phosphorus, and she'd find it and throw it away."

After an explosion of pyrophoric red phosphorus sent David to the hospital and almost blinded him, Kathy and Ken laid down the law: No more hands-on chemistry of any kind in the house. David began spending more time at his mother's house in Commerce Township, where Patty Hahn lived with her boyfriend, Michael Polasek, a retired General Motors forklift operator. David set up a lab in his mother's potting shed and began working on a new project—one that would take him well beyond the realm of conventional chemistry.

Patty and Michael thought it strange that David spent so much time in the shed and that he would sometimes wear a gas mask and discard his clothing after coming out. When they asked what he was up to, David provided explanations that, as Michael said, "went right over my head." One thing Michael remembered, though, was that David said his experiments had something to do with "creating energy."

The very idea of Taylor keeping a secret about anything he's passionate about is inconceivable to anyone who knows him. The ten-year-old Taylor threw himself into the study of atomic energy with characteristic intensity and talked nonstop about his new passion. He maxed out the local library's thin collection of books on nuclear physics and history, which presented nuclear science as a rarefied world inhabited by PhDs and Nobel laureates whose brainstorms were made possible by multibillion-dollar machines and laboratories. Hahn's story gave Taylor a glimpse into a world where hands-on nuclear physics was something that an individual—a kid, even—could try.

"The nuclear spark really hit me when I read that book," Taylor says. But Taylor didn't want to copy David Hahn. In fact, he could see exactly where Hahn had gone wrong. Taylor questioned whether Hahn could create a neutron source with the methods he'd used and the quantities of materials he was able to collect. And he didn't think Hahn came close to breeding uranium-233—"although," says Taylor, "he did a pretty good job contaminating the place."

What was most intriguing to Taylor was that an amateur had gotten involved in atomic science: "I thought, assuming I do it smarter than him, I could actually play with this stuff. I could do what he was trying to do—but I could be the responsible radioactive Boy Scout."

The upcoming fifth-grade science fair gave Taylor an opportunity to ease into hands-on nuclear science. "I knew my parents wouldn't be gung-ho to get into radioactive stuff to begin with," he says. And so he proposed to assemble a relatively benign "survey of everyday radioactive materials" for the fair.

Natural radioactive decay, as Taylor would tell the parents, teachers, and students who stopped by his display, is the spontaneous emission of high-energy particles from radioactive materials. Radioactive elements spend their lives destroying themselves, their atoms releasing energy from their nuclei as they decay.

All matter is made of combinations of elements—substances that can't be further broken down or transformed into other substances by chemical means. Taylor had, by this time, studied the periodic table well enough to know which elements were radioactive. The periodic table has been chemistry's primary map since the mid-1800s. Frequently expanded, refined, and debated, it classifies elements according to chemical properties, electron configurations, and atomic numbers—the last referring to the number of protons found in the nucleus of an individual atom.

An atom is the smallest particle of an element that still has that element's distinctive chemical properties. The word *atom*—from the Greek word *atomos,* "indivisible"—means "that which cannot be divided further into separate parts," though scientists now know that atoms themselves are made up of even smaller particles. At the center of every atom is the nucleus, a tight, positively charged ball of protons and neutrons that is surrounded by negatively charged electrons.

Atoms of so-called light elements, such as hydrogen or helium, have few protons and neutrons in their nuclei. These atoms, which were created during the first moments of the universe, are stable and even-tempered, capable of crisscrossing the cosmos intact or hunkering down in the middle of a buried rock while civilizations rise and fall and stars spark to life and die. But heavier atoms—those with a surfeit of protons and neutrons in their nuclei—are more jittery; they spend their lives throwing off unwanted energetic particles as they seek stability. The emission of those particles is called radiation.

Taylor had become fascinated by the heavy elements residing in the neighborhoods near the bottom of the elemental map. Particularly beguiling to him were fifteen elements in the lowest row, a group known as the actinides. These metals are all extremely unstable. Uranium, which in its most common form has 238 protons and neutrons stuffed into its nucleus, is one of the heaviest naturally occurring elements. As it decays, striving to attain balance, it spits out chunks consisting of two protons and two neutrons, a combination known as an alpha particle. As it loses alpha particles, it transforms into different elements, eventually arriving, after 4.47 billion years, at the fourteenth and final stage of its decay chain, lead-206, at last achieving its long-sought stability.

After Hahn's father and stepmother forbade David's scientific endeavors, the angry teenager pressed on in secret. Kenneth and Tiffany thought of steering their son toward more benign pursuits but they resisted the impulse, which couldn't have been easy given that their child, who had a talent and fondness for blowing things up, was now proposing to dabble in nukes. Kenneth borrowed a Geiger counter from an acquaintance at Texarkana's emergency-management agency and over the next few weekends, he and Tiffany shuttled their son around to nearby antiques stores, where Taylor pointed the clicking detector at old radium-dial alarm clocks, thorium lantern mantles, and orange uranium-glazed Fiesta plates.

Tiffany and Kenneth have a picture of Taylor wearing glasses, braces, and his school shirt standing next to a poster titled "Radioactivity of Everyday Products." The narratives under the subheadings—Problem, Hypothesis, Procedure, Applications, Materials, Data, Results, Conclusion—describe the experiments, which involved measuring the radioactivity of the plates (uranium oxide), clocks (radium), lantern mantles (thorium), and salt substitute (potassium-40).

The science-fair project was a hit. But Taylor wasn't content to collect a few radioactive plates. Drawn in by what he calls "the surprise properties" of radioactive materials, he needed to know more. Taylor's knowledge of chemistry had allowed him to fuel his home-recipe propulsion program and grasp the underpinnings of his biology and genetics experiments. But both chemistry and biology are built on an understanding of the laws of physics. Taylor started to suspect that atoms, so small but so energetic — much like the young Taylor himself — could offer a lifetime of powerful secrets to unlock.

With atomic physics, Taylor had finally met something worthy of his restless intellect. He would continue to play his pyrotechnic parlor games each Fourth of July, but his giant rocket would go unfinished, and the biology experiments under way in his grandmother's garage would remain half done. He became obsessed with nuclear energy, soaking up every particle of knowledge he could find. Just as he had taken apart his rockets to see what made them fly, he wanted to break down the basic units of matter, to understand how the universe works at its most fundamental level.

After Kenneth returned the borrowed Geiger counter, Taylor used some of his singing earnings to purchase one of his own. In his grandmother's garage, on tables already crowded with chemicals and microscopes and germicidal black lights, an expanding collection of radioactive objects began to appear: vacuum tubes, old glow-in-the-dark wristwatches and aircraft gauges, and an array of antique figurines, clocks, and gadgets. Taylor also bought a pig — a radiation-shielding container made of thick lead — to stash the most radioactive materials in.

"I think the draw of nukes," says Tiffany, "was that it was something more complex than anything he'd been doing; it was a big challenge."

There was also the forbidden-fruit aspect, as compelling to Taylor as it had been to David Hahn. But whereas Hahn had had to keep his experiments secret, Taylor was delighted to find that he became the center of attention whenever he talked about the "radioactive stuff" he was collecting and experimenting with. He had learned to catalyze awe and adulation first with his singing, then with his explosive chemistry. Nuclear energy offered an even bigger stage with the more substantial imagery of power, intrigue, creation, and destruction. Even in his linebacker-worshipping corner of the heartland, a kid who could wrap his head around the mind-blowing mysteries at the base of all matter could command attention and respect. He could also, potentially, put on one hell of a magic show.

PART II

6

THE COOKIE JAR

"DOES ANYONE KNOW who Marie Curie was?"

Taylor, in sixth grade now, wears his lab coat, bright purple rubber gloves, and a surgeon's mask pulled down around his neck. He circles a table topped with gadgets, many labeled with the yellow-and-black radiation symbol known as a trefoil. He locates two black-and-white pictures and holds them high for the class to see. "These are the Curies, Marie and Pierre. I'm sure at least one person in here knows who Marie Curie was?"

As Kenneth pans the camcorder around the classroom, no faces illuminate with recognition.

"I'm *ecstatic* about her!" Taylor shouts, thrusting his spindly arms into a Y shape, comically accentuated by the purple gloves. "Marie Curie is the most amazing woman I have ever met—well, I've never met her because she's dead, you know. But she practically *discovered* radioactivity!"

Actually, Marie Curie coined the term *radioactivity,* in 1898, for the phenomenon described by French physicist Henri Becquerel two years earlier. Curie's research into the extraordinary properties of radioactive materials overturned established ideas in chemistry and forced scientists to reconsider the foundations of physics. As a result, she became the first woman to win a Nobel Prize (for Physics, in 1903; she shared it with her husband, Pierre, and Becquerel). Taylor, who began his talk by drawing a classic picture of an atom—a nucleus of protons and neutrons circled by electrons with streaking comets' tails—tells the class about Marie Curie's discovery of radium, the most intensely radioactive of all the relatively accessible naturally occurring elements. "Radium," he says, "is what killed her."

He pauses dramatically as he opens the top of a round lead jar. "And now," he says, "I'm going to show you some real, purified radium."

Marie Curie died from aplastic anemia, which is associated with exposure to ionizing radiation. Curie and her collaborators had little sense of the damage that radiation could do to human tissue. They worked unprotected for decades; Marie carried radioisotopes in her pocket, stored them in her desk drawer, and marveled at the faint blue glow emanating from her test tubes:

> One of our joys was to go into our workroom at night; we then perceived on all sides the feebly luminous silhouettes of the bottles or capsules containing our products. It was really a lovely sight and one always new to us. The glowing tubes looked like faint, fairy lights.

Curie developed mobile x-ray stations to diagnose wounded soldiers in the First World War, then spent long hours operating her unshielded machines in wartime field hospitals. There was so much radiation bouncing around Curie's laboratory and life that her logbooks from the 1890s, and even her cookbook, still retain enough radioactivity that they have to be stored in locked lead-lined boxes. Researchers (who must wear protective clothing) can see the indelible fingerprints that her radioactive fingertips left on the photographic film tucked between pages.

Taylor has been talking nonstop for forty-five minutes while his assistant, Noah Jackson, stands patiently behind the table. Noah, wearing a mask and gloves, awaits instructions as Taylor, like a magician pulling rabbits from a hat, plucks one radioactive goodie after another from his lead pigs. Apart from asking Taylor to pull his mask down from over his mouth—"So we can all hear you better"—teacher Angela Melde has let her most intellectually precocious student keep rolling with his peculiar show-and-tell.

In the three years since his astronaut presentation, Taylor has grown intellectually and physically. But emotionally, Taylor the eleven-year-old does not seem to be much more mature than Taylor the nine-year-old. Though many of his classmates are moving into the self-censoring coolness of puberty, Taylor lacks any hint of a typical sixth-grader's self-consciousness. He prances around Peter Pan–style, flailing his purple-gloved hands as he proclaims the wonders of nuclear power, the electromagnetic spectrum, and cosmic radiation. When he pulls out plastic-encased samples of radioactive cesium and strontium, he clutches them to his heart and rocks back and forth, converting his voice into a baby's squeak: "These little things, I like to call them my little pets, these radioactive friends, because they're so cute. I love them so much."

"What are their names?" a kid asks.

"I call this one Cesi, and this one Stronti," Taylor replies. He hands the small disks to a girl in the front row to pass around. The girl hesitates a moment before taking it.

"They're safe?" Mrs. Melde asks.

"Well, you don't want to put them in your bed with you tonight," Taylor replies, drawing the class into laughter, "but for a short time, they're okay."

The class watches as Taylor, taking his time, pulls a larger, thicker orange glove over his right-hand purple glove. "These are my contamination gloves," he says, "for when I'm working with very hazardous things. I'm going to keep them on, because I'm scared of these things I'm going to show you now."

Taylor lifts his mask over his face and reaches into the jar with his orange-gloved hand. He pulls out a dial, salvaged from an early Cold War–era aircraft, painted with radium-226. "You can see how hot it is," Taylor says, picking up a Geiger counter. "Actually, you can *hear* it. I wonder how many of those pilots got cancer." Having come to associate danger level with the intensity of the sounds coming from Taylor's radiation detector, the students gasp at the frenzy of beeps.

"I'm going to put it back behind the lead," Taylor says, dropping the dial into the cookie jar, "where it will be safe." He demonstrates by applying the instrument to the outside of the jar; it doesn't click at all.

Kenneth, from behind the camcorder, chimes in. "Taylor, tell 'em *why* they painted radium on the dials and numbers."

"It glows in the dark!" Taylor says. "There was a radium craze in the twenties, and they were putting radium-226 on airplane instruments, clock faces, car dials, everything. If it was darker in here, we'd see it glowing." He pulls an old clock out of the jar and tells the class that, even though he washed most of the radium paint off the numbers, the clock is "still radioactive, and it will be for more than a thousand years."

A student asks why he washed it off, and Taylor's face lights up.

"Ah! Well, what I wanted to do in my laboratory at home is I wanted to purify radium and make it even more radioactive so I could use it to bombard stable things like nickel and transform it into an unstable isotope like nickel-65, which is radioactive." He digs into the jar again and pulls out a vial. "Look down in this tube and you'll see a white substance. That's actual radium that I've collected and purified. It's some of the most dangerous stuff on earth."

Taylor playfully waves the small glass ampule at a girl in the front row, who ducks.

Among the elements on Taylor's periodic table, few are more frightening than the one with the atomic number 88. Radium atoms emit radiation at an intensity three thousand times greater than an equal number of atoms of uranium. Outside the body, radium's alpha-particle emissions are easily blocked. But if radium gets inside the body, its effects are devastating.

Radium has become something of a fetish for Taylor the sixth-grader. Hahn, too, became obsessed with radium, though at a later age. He'd caught the nuclear bug while completing the requirements for the Boy Scouts' Atomic Energy merit badge, and he decided to build a "neutron gun" to bombard elements with neutrons and change their atomic structure. "My goal," he told me, "was to irradiate every element on the periodic table and see what would happen." Hahn took apart more than a hundred smoke detectors and combined the tiny bits of radioactive americium-241 he'd found inside each one with lithium. But Hahn found that the neutron ouput wasn't nearly enough to produce protactinium-233, which decays into fissile uranium 233. He needed something stronger.

When Hahn got his driver's license at sixteen, he began scouring antiques stores and junkyards for objects that might contain radium. Unlike Taylor, he didn't bring his parents along. Fearful that they'd try to shut down his collecting and experimenting, he increasingly kept his scientific pursuits to himself. When Hahn found a radium clock or instrument, he'd bring it to his potting-shed laboratory, scrape the paint off with a knife, and collect the flakes in small containers.

Later, I would ask Taylor if he did the same thing when he was eleven years old and working with radium. "No!" he blurted out. "Scraping is the last thing you'd want to do. When you scrape, you get dust floating around, and you could breathe it in. Radium is a bone-seeker; it's terrifying what happens when radium gets inside you."

After penetrating the membrane linings of the lungs or gastrointestinal tract, radium is absorbed into the bloodstream, where it works its way to the bone marrow. Once deposited in the marrow, it becomes a permanent internal radiation source, bombarding bone- and blood-forming cells with alpha radiation. Since radium has a half-life of 1,601 years, it is there to stay for the rest of a person's — likely shortened — life.

"I went to incredible precautions," Taylor tells me. He dabbed acetone on the radium-coated surfaces, then decanted off the paint and let the acetone evaporate out, leaving mostly radium salts. As he worked, he set up several Geiger counters and wore a finger dosimeter, a ring that measures exposure to ionizing radiation.

Hahn, working with inconsistent safety protocols — sometimes he wore a gas mask and gloves, sometimes not — never allowed others to witness his experiments. For the sake of anyone who might have dropped by the poorly ventilated shed, that's probably a good thing.

Taylor tops off his classroom presentation by demonstrating how an isotope generator works. It's a classic and once-common university nuclear physics experiment (though it's rarely done now, due to concerns about danger and expense), but the kids at St. James Day School are getting it in sixth grade from an eleven-year-old teacher.

"Every isotope has a half-life. That's how long until it takes for half the radioactive nuclei to decay. So if uranium-238 has a half-life of two billion years" — it's actually 4.47 billion years, about the same age as the Earth — "half the radiation will be gone in two billion years. When doctors inject something into your body, you don't want it to emit radiation all your life. You want it to show you the cancer and be gone."

Taylor says that nuclear pharmacists use isotope generators to "milk" technetium-99m — nuclear medicine's most widely used tracer— from its parent isotope source, molybdenum-99. But since molybdenum-99 is more hazardous, Taylor is using cesium-137 as a stand-in.

Taylor's friend Ellen Orr volunteers to put a few drops of saline (which Taylor tests with his detector to show that it's not radioactive) into the generator. As it flows down through the cylinder, Taylor explains that the liquid is washing out barium-137m atoms that come from beta decay of the cesium-137. The *m* in barium-137m, Taylor explains, stands for *metastable.* That means it has an excited nucleus that will quickly decay to a stable isotope.

"We nuclear scientists like to joke about parents and daughters and cows and pigs," Taylor says. "Think of cesium-137 as the parent, and the cow is the isotope generator. You milk out the daughter, barium-137m, which you put in the pig."

He moves his Geiger counter's probe close to the liquid that drips into the bottom of the container, and it clicks like crazy. The whole class starts laughing and cheering.

"Isn't that neat?" Taylor says. "What you put in isn't radioactive, but what comes out is! I'm going to put a lid on this, and I'll pass it around with the Geiger counter." He explains that barium-137m has a half-life of less than three minutes, and that within thirty minutes it will have decayed to less than one-thousandth of its initial activity, making it safe to dispose of.

"But till then, y'all, don't take that lid off!"

Taylor had gotten his crowd-pleasing isotope generator just before his eleventh birthday during a trip he and Kenneth took to Oak Ridge, Tennessee, the wartime Secret City whose factories produced enriched uranium for the first atomic weapons. After he and Kenneth visited the museum, Taylor asked his dad to take him to Spectrum Techniques, a supplier of nuclear-research equipment. "I'd already collected a bunch of radioactive stuff, collectibles from eBay and stuff," Taylor says. "Spectrum had more serious stuff, not really for hobbyists, more for the nuclear industry."

The ten-year-old rehearsed his approach with Kenneth on the way over. When they arrived, Taylor walked up to the receptionist. With his chin barely clearing the countertop, he announced, "We're interested in purchasing some small check-source disks. Maybe cesium-137 or strontium-90, if you have them."

"The guy's jaw kinda dropped," Kenneth says. "For a few seconds he couldn't say anything. Then he got up and said, 'Uh, let me get a sales associate for you.'"

The sales associate was just as gobsmacked. Taylor introduced himself and said he was an amateur nuclear scientist. The associate asked why he wanted the sources, and Taylor told him about his experiments.

"At first he was just amazed, and then we ended up having quite a conversation," Taylor remembers. "Then he said, 'Hold on a second, Taylor,' and he went in back." He came out with two small, brightly colored disks, each an inch in diameter: one-fifth of a microcurie of cesium-137, and one-tenth of a microcurie of strontium-90. The sealed check sources typically sold for between fifty and a hundred dollars.

"These are on us," the sales associate said.

"This is awesome!" Taylor said. Then he told the associate that he'd *really* love to have a cow, an isotope generator. Unfortunately, the associate said, the margins were too low to give away one of the three-hundred-dollar devices.

"Well, Dad, what do you think, could we just buy one?" Taylor asked.

Kenneth hesitated. It was already an expensive trip. After a few awkward moments, the associate motioned for Taylor and Kenneth to follow him. He led them through the production facility and into the warehouse, where he scanned a high shelf, found a box, and pulled it down.

"This one here's a demonstration unit," he said. "It works like new." He handed it to Taylor. "It's yours."

"Are you sure?" Kenneth asked.

The associate looked at Taylor. "One of these days, I get the feeling you'll repay us."

"I decided that no matter where my career went," says Taylor, "I would always go back there to buy stuff and I'd refer everyone to them —which I have."

Taylor and Kenneth also stopped by the University of Tennessee's campus at Oak Ridge National Laboratory to meet Lee Dodds, chair of the university's Department of Nuclear Engineering. One of Kenneth's nephews had arranged the meeting, but Dodds had somehow gotten the impression that he'd be talking to a potential graduate student; he was shocked when tiny Taylor walked in. Dodds's secretary had scheduled thirty minutes, but the esteemed professor immediately looked at his watch and said, "I've got ten minutes."

Three hours later, Dodds and Taylor were still together, laughing and talking nukes. "He was by far the smartest and most advanced ten-year-old I'd ever met," Dodds remembers. "But he wasn't a nerd or bookworm; he was extroverted, laughing and joking. He knew more than some PhD students, and he made more eye contact than most."

Dodds took Taylor and Kenneth on a tour of the labs and introduced Taylor to faculty members. At the end of the day, Dodds told Taylor, "Our department has its own scholarship for the most promising nuclear engineering students. If you do well on the ACT [college admissions test], that scholarship is waiting for you."

On the return flight, Kenneth recalls, "All I could think about was getting home and telling Tiffany that our ten-year-old had just been offered a college scholarship."

7

IN THE (GLOWING) FOOTSTEPS OF GIANTS

ANGELA MELDE HAD GIVEN her students a maximum of thirty minutes for their presentations. But an hour after he'd begun, Taylor was still going on about the Manhattan Project, about dark matter and string theory, about Ernest Rutherford and Enrico Fermi. Why did his teacher let him continue?

Melde had recognized that Taylor learned best by talking. "There aren't that many people who are verbal learners to his extreme," Melde says, "but for some, the concepts really jell and synthesize when they have a chance to explain them. When a teacher sees that much enthusiasm and that much engagement with a subject coming out, it would be a crime to stop it, because something magical is happening in terms of learning."

Only after ninety minutes did Melde finally ask Taylor to wrap it up.

"I guess," Taylor muttered, "we won't be able to get into quantum mechanics."

Neither Melde nor head of school Dee Miller had any formal training in educating gifted children, but St. James was committed to providing an accepting culture and an individualized learning environment for each student. "We have kids with handicaps and kids who are really talented in one way or another, kids who might be considered weird elsewhere," says Miller. "We work hard at tuning in to what each boy or girl needs and creating a sort of micro-environment for them that helps them learn best."

One afternoon after school let out, Taylor stopped by Dee Miller's small second-floor office. The school had recently purchased new computers, and Taylor wanted to know if he could have the old CRT (cathode-ray tube) monitors now stacked in a corner of the computer lab. He told Miller that he wanted to open them up and "try some experiments."

Miller and the teachers had always made things available to Taylor, but this worried her.

"Taylor, I don't open up monitors; they can be dangerous."

"Well, I need 'em for my lab," Taylor said.

"Tell you what," Miller said. "You get your daddy to come up and tell me you need them and then we'll consider it."

Taylor brought Kenneth in the next afternoon, but, says Miller, Taylor did most of the talking. They drove away with three monitors, which Taylor promptly tore apart when he got them into his grandmother's garage. A couple of days later, Tiffany snapped a picture of her smiling son surrounded by lead shielding and the spilled-open guts of the monitors, one of the bare CRTs wired to a high-voltage supply.

As Dodds had recognized, Taylor was more than a bookworm. His curiosity drove him toward both theory and hands-on creation. If he read or heard about a phenomenon or a discovery that grabbed his interest, he'd immediately start thinking about ways to test it or build something practical to do with it. Then he'd head to his laboratory and start tinkering.

Taylor wanted to use the CRTs to reproduce the discovery that German physicist Wilhelm Konrad Roentgen made on November 8, 1895. Much of the debate among physicists in the second half of the nineteenth century revolved around whether light, heat, and other radiation were all part of a continuum. By the 1880s, scientists were feeling more confident about their understanding of what would come to be known as the electromagnetic spectrum. On the long-wavelength end of the spectrum are radio waves, preceded by shorter microwaves, which are preceded by even shorter infrared waves. Visible light, thought to be near the shorter-wavelength end of the spectrum, occupies a narrow range of frequencies before infrared. But the existence of wavelengths shorter than ultraviolet light was unknown.

Soon, a series of discoveries would collapse the foundations of late-nineteenth-century physics. The first occurred when Roentgen was experimenting with cathode rays, created by passing high-voltage current through an evacuated glass container. Cathode rays would eventually form the basis of every television and computer screen until liquid-crystal displays (LCDs) displaced them, a century after Roentgen's experiments.

Roentgen was investigating the properties of cathode electron beams, which are very short range, less than an inch. Knowing this, the German was surprised when he switched on his electron beam and noticed that some fluorescent material suddenly glowed on the other side of his laboratory—even though the end of the tube was covered by cardboard. Roentgen then set up a screen and held his hand between it and the tube. In what must have been a moment worthy of a thousand shivers, he saw an image of his finger bones projected onto the screen.

Roentgen had discovered x-rays.

Six weeks later, the scientist incorporated film into his experiment and made the first radiograph, the now-famous x-ray image of his wife's hand—bones, wedding ring, and all. Roentgen's feat was quickly copied, and x-ray images became a worldwide sensation, celebrated variously as a technological attraction, an art form, and a magic visual gateway to the supernatural world. But it didn't take long for Roentgen's new form of electromagnetic radiation to prove its value in medicine, astronomy, security, and many diverse areas of science.

Taylor was fascinated by the possibilities of accelerating electrons through a vacuum. "The tubes in these computer monitors are a lot like the fluoroscopes in the old x-ray machines," he says as he shows me his crude but functional x-ray devices. "You've got a power source that kicks up tens of thousands of volts and accelerates negatively charged electrons through a funnel-shaped vacuum tube. The x-rays get thrown up against this glass plate, which is doped with fluorescent or phosphorescent compounds, and the electron beam creates the image. The glass is leaded, so most of the x-rays stay inside." Taylor tried different configurations for his x-ray machines using the school's CRTs, rectifier vacuum tubes from old radios, even an ornate antique Crookes tube similar to what Roentgen had used.

As he was showing me the remnants of his experiments, it occurred to me that Taylor's preternaturally early progression of experimentation was following—and would continue to follow—roughly the same path that particle physics itself had: from the discovery of x-rays and radioactive compounds to the use of alpha particles to induce nuclear reactions to the development of electrostatic accelerators and artificial radioactivity to, finally, nuclear fission and fusion.

Inspired by Roentgen, Henri Becquerel began researching x-rays, theorizing that they were caused by phosphorescent compounds, which he believed were influenced by sunlight. One by one, he placed his specimens on photographic plates wrapped in black paper and set them in the sun, assuming that the x-rays would pass through the paper and expose the plates with their telltale signatures. But the day in 1896 on which he'd planned to test uranium salt crystals dawned cloudy, so Becquerel stashed both the uranium and the photographic plate in a drawer. Several days later he processed that plate and discovered that the uranium salts had left an image. Somehow, the uranium was spontaneously emitting some sort of penetrating radiation that had nothing to do with sunlight.

"I did the same thing," Taylor says. He'd borrowed some film from Tiffany's camera and placed samples of various radioactive materials on top of it — uranium ore, radium gauge hands, thorium lantern mantles. "The film I had wasn't the best for the job and my technique was fuzzy, but when I developed it, sure enough, there were pictures of the stuff I put there."

But x-rays weren't responsible for exposing Becquerel's or Taylor's film. Becquerel had stumbled upon yet another class of energy, a high-frequency electromagnetic wave within the larger phenomenon that Marie Curie would call radioactivity. But this energy didn't emanate from atoms' electron clouds. Scientists would eventually learn that this energy — gamma rays, which have the smallest wavelengths and the most energy of any wave in the electromagnetic spectrum — is emitted from excited atomic nuclei. Due to their short wavelength, gamma rays are the most penetrating form of radiation and, like x-rays, can travel great distances through air.

As Taylor was ramping up the scale and ambition of his experiments, he was also becoming an ever more obsessive collector. After dinner he'd typically skip his homework and spend the evening scouring eBay, seeking out nuclear-related items. Some he bought to experiment with; others he bought just because he thought they were cool. Some became part of his sixth-grade science-fair project, an expanded study of environmental radioactivity that included radioactive samples gleaned from nuclear power plants and nuclear test sites. Taylor built a glove box of wood and Plexiglas with a large PVC pipe that could be unscrewed from outside or inside the chamber — "just like in the movie *The China Syndrome*!" he excitedly told those who stopped to view his display. He bought a more sophisticated sodium iodide scintillation detector and invited visitors to reach into the gloves and handle the materials (which had radiation levels low enough that they could have been safely picked up and held—for a short time, at least—with bare hands).

Taylor used his school computer-lab time to study uranium chemistry. "I knew he wasn't doing what I'd assigned," says Melde. "But I also knew he was learning a lot, so I let him do his own thing." Likewise with recess: Taylor got along with everyone, Melde says, but in the sixth grade, he stopped playing with his classmates. While the other kids chased one another and played soccer, Taylor would sit on a bench with Melde and go through a binder filled with correspondence from physicists and scholarly articles about how uranium processors transformed weakly radiating rocks into reactor-grade fuel that is, ounce for ounce, more than two million times more energetic than coal.

Uranium is produced by supernovae, which occur when stars burn through their nuclear fuel. The star's core collapses catastrophically and brightens in an explosion that boosts its luminosity by a factor of ten billion, outshining the entire galaxy that contains it. Inspired by Becquerel's discovery, the Curies experimented with techniques to extract uranium from an ore called pitchblende. Enriched uranium is the chain-reacting stuff of atomic bombs and power stations, but the element in its raw form decays so slowly in its multibillion-year-long cascade toward stability that uranium ore is only slightly radioactive.

After separating the uranium metal from the ore, the Curies were surprised to find that the leftover waste products were *more* radioactive than the extracted uranium. To explore the phenomenon, Marie acquired tons of pitchblende tailings from a factory in Austria and experimented with the material until she isolated two previously unknown, highly radioactive elements: polonium and radium. In 1911 she won a second Nobel Prize (for Chemistry) in recognition of these discoveries, making her the only person to win a Nobel in two different sciences.

One day at the school computer, Taylor came across a description of a process that extracts uranium from ore. He realized that it was something he might be able to do at home. That night, he took a heavy hammer and began whacking at a chunk of uranium ore he'd purchased on the Internet. After a night of experimenting with different combinations of chemicals and heating methods, he went out to Grandma's garage, "and there was this beautiful yellow fluorescent solution" (uranyl acetate).

Next, he tried leaching concentrated pitchblende (uranium dioxide) with acetic acid and hydrochloric acid. He filtered the leachate with a vacuum flask and let the solution evaporate. "Those uranium salts really took off and grew some incredibly beautiful yellow fluorescent crystals. That was cool! I tried different combinations of chemicals and it got even better. The colors were just amazing!"

If Taylor had a true counterpart at the dawn of the nuclear era, it was the English physicist Ernest Rutherford. Enthusiastic and relentlessly curious, Rutherford was fascinated by the esoteric powers of the atom. At the time, most physicists believed that positive and negative charges were distributed evenly inside atoms. But Rutherford, in his most famous experiment, proved otherwise. In 1911, he shot a narrow beam of alpha particles emitted by radium at a thin sheet of gold foil. Most of the particles passed through the foil onto the screen behind it, but some of the particles bounced back, deflected by the gold.

"It was as if you had fired a 15-inch naval shell at a piece of tissue paper and the shell came right back and hit you," wrote Rutherford. "It was then that I had the idea of an atom with a minute massive centre carrying a charge."

Rutherford realized that a few of the alpha particles were colliding with something so dense they were deflected away, which disproved the theory that particles simply traveled through atom "clouds." Rutherford concluded that what they were encountering was the central part of the atom, which held most of the atom's mass; he dubbed this small, dense concentration of charge and mass the *nucleus,* from the Latin word for "kernel." Two years later, Hans Geiger, a collaborator on the gold-foil experiment, would verify these conclusions and prove Rutherford's theory. Thereafter, studies of the nucleus became known as nuclear physics, and they would lead to the discovery of the proton and neutron (by Rutherford and his research partner James Chadwick) and eventually to the splitting of the atom.

"Those were the discoveries," says Taylor, "that made my career possible."

Among the enablers of Taylor's budding career, few were as useful as the Internet. Little more than a decade earlier, the still-infant web had yet to connect far-flung amateurs or provide easily accessible sources of information on their esoteric interests. To get guidance on techniques and find sources for the radioactive materials he sought, David Hahn posed as a physics instructor and initiated snail-mail correspondences with officials at organizations listed in his Boy Scout merit-badge booklet. Amazingly, no group proved more helpful to Hahn's endeavors than the Nuclear Regulatory Commission (NRC) itself. Responding to Hahn's inquiries, Donald Erb, the NRC's director of isotope production and distribution, sent "Professor Hahn" tips on isolating radioactive elements. He also provided a list of isotopes that could sustain a chain reaction, as well as commercial sources for the radioactive wares Hahn wanted to purchase.

Taylor started nosing around online in 2006 or so, and he quickly made connections with hobbyists who were eager to share their knowledge and point him toward sources for radioactive materials. To finance his purchases, he started a small business buying broken iPods, fixing them, and reselling them on eBay. Guided by his similarly obsessed cyberfriends (most of whom had no idea they were communicating with an eleven-year-old), Taylor expanded his collection of radioactive "naughties," as some in the community refer to them.

"You'd actually be surprised at some of the things you can buy online," Taylor says. *Unsettled* might be a better word. While researching and writing this book, I often looked up terms like *minimum critical mass of enriched uranium,* and I sometimes wondered if I was raising an NSA data-miner's eyebrows, but Taylor's nightly cyberhunts and purchases attracted no inquiries from regulatory or law enforcement agencies. Considering the degree of official paranoia focused on more benign materials and information-gathering since 9/11, that's surprising. Considering that Taylor was an eleven-year-old child when he began buying radioactive materials on the Internet, it's disturbing.

But Taylor undoubtedly knew more about the proper handling of these materials than many of the online buyers and sellers—nearly all of whom, it's safe to presume, were older than him. Unfortunately, some apparently were not old enough to know better, as evidenced by the thirty-one-year-old Swedish do-it-yourselfer who in 2011 tried to build a nuclear reactor in his kitchen using materials he'd bought on eBay. Richard Handl started cooking a mix of americium, beryllium, and radium in a broth of sulfuric acid in a confused attempt to "see if it's possible to split atoms at home." Though his approach wasn't capable of splitting atoms, he did achieve a meltdown of sorts when the materials exploded in his face. Handl then called Sweden's Radiation Authority and asked if what if he was doing was legal. It was not. The authority directed police to Handl, who was charged with unauthorized possession of nuclear material.

Taylor bought more lead pigs to contain his newest treasures: samples from nuclear power plants; check sources from nuclear weapons tests; chunks of thorite (the ore source of thorium); radioactive glass from mortar sights and other weapons used by American, Soviet, and Chinese forces. Some of these would play a part in Taylor's increasingly ambitious experiments. Others, such as his Revigator—an earthenware drinking-water crock lined with radium-226 or other radioactive materials—were bound for his collection of antique radioactive quack cures.

Some of Taylor's artifacts containing significant quantities of radium-226 are as dangerous now as they were when they were created, in the days before anyone fully understood the damage that ionizing radiation could do. They are also, counterintuitively, some of the least controlled radioactive items in the U.S. Like antique cars exempt from emissions standards, these materials are not inspected or tracked. They're regulated under a "general license" from the NRC that allows "any person to acquire, receive, possess, use, or transfer" them, as long as the materials are not dangerously altered or damaged in a way that could result in a loss of radioactive material. In that case, the owner is required to notify the NRC.

8

ALPHA, BETA, GAMMA

THE RADIOACTIVE QUACK-CURE craze began in the early twentieth century after British physicist J. J. Thomson, a Nobel laureate who had discovered the electron, wrote a letter to the scientific journal *Nature* in which he announced that he had found radioactive gas in well water. This inspired others to check the waters in many of the world's most famous health springs; several were found to have elevated levels of radioactivity. In one of history's more catastrophic breakdowns of logic, medical experts jumped to the conclusion that radioactivity must be responsible for the springs' supposedly therapeutic properties. U.S. Army surgeon general George H. Torney wrote glowingly of the relief that radioactive water could bring for ailments from gout to malaria to chronic diarrhea. Dr. C. G. Davis noted in the *American Journal of Clinical Medicine* that "radioactivity prevents insanity, rouses noble emotions, retards old age, and creates a splendid youthful joyous life."

Suddenly, radiation was all the rage. In Europe and North America, health- and status-seekers drank Radithor, a brand of radium-treated bottled water. Radium-infused ointments, toothpastes, contraceptives, and suppositories became the latest craze, and uranium-containing cigarette holders were marketed as a means to fend off cancer. The wackiest invention of them all may have been the Radiendocrinator, a sort of radium-soaked jockstrap that men were advised to place under their scrota at night to stimulate endocrine glands and boost virility.

The negative health effects of radiation exposure didn't become well known until the death of Radithor's most famous victim, Eben McBurney Byers, in 1932. The wealthy industrialist drank more than a thousand bottles of the stuff over three years; most of his jaw fell off, and holes opened up in his skull. He was buried in a lead-lined coffin. Byers's widely reported demise boosted public awareness of radiation's dangers and led the U.S. Congress to strengthen the Food and Drug Administration's enforcement powers. But the inventor of Radithor and the Radiendocrinator, William J. Bailey, demonstrated his faith in his products by continuing to use them, telling people he'd "drunk more radium water than any living man," until he succumbed to bladder cancer.

During the height of the radiation quack-cure craze, if you wanted to produce your own radioactive water, you could buy a Revigator, patented in 1912 and marketed as a remedy for arthritis, flatulence, and senility. Though he had no intention of using it for its intended purpose, Taylor managed to find and purchase one of the uranium-ore-lined ceramic crocks with a spigot at the bottom to dispense the irradiated water. He stored it in his grandma's garage amid the expanding jumble of gadgets, vials, lead pigs, and instruments.

Each time Kenneth, Tiffany, or Nell visited the garage, there'd be something new: a pre–Geiger counter radio-assay electroscope; a radioactive Chinese army mortar sight; a thoriated camera lens; a check source for ion-chamber instruments used at the Nevada Test Site. Taylor bought and tacked up an official-looking yellow sign reading CAUTION: RADIOACTIVE MATERIALS. He exchanged his white lab coat for the canary yellow favored by nuclear technicians, and he purchased a yellow-orange Tyvek hazmat jumpsuit, which he hung from a nail on a rafter—"just in case," he told his parents, "of any unusual incidents that might involve contamination."

Since the rocket-candy incident, Taylor had gotten increasingly adamant about safety—at least, from what his parents could tell. "It wasn't just the facemask anymore," Tiffany says. "He started wearing his gloves and respirator a lot too, and setting up all kinds of instruments."

For Tiffany and Kenneth, the silent, invisible creep of radiation was more unsettling than any of the explosions of Taylor's pyrotechnic era. As bone-jarring as those could be, at least a parent could see and hear them. "What scared me," Tiffany says, "was that you couldn't tell by looking at any of this new stuff how dangerous it was. Taylor could explain it all, but I didn't get most of it. Honestly, we didn't know if it was safe or not."

Indeed, unless you were one of the few people who could recognize a Revigator, you'd never know what kind of damage this dapper, flapper-era thing was capable of. Likewise with the strange and ever-multiplying cache of rocks and metals and vials of liquid that increasingly crowded Taylor's laboratory, some of which cast a glow into the corners of the garage at night.

"I had a lot of faith in Taylor, and he seemed to know what he was doing," Kenneth says. "But when he started bringing home more stuff and explaining what it was, we were getting pretty worried."

Near the end of Taylor's visit with Lee Dodds at the University of Tennessee in Oak Ridge, Kenneth had confided to the professor that he had concerns about his son's safety. Dodds, who'd spent the past three hours talking nuclear science with Taylor, looked Kenneth in the eye and said: "He knows what he's doing."

After Dodds's endorsement, Taylor ramped up his collecting and experimenting. Unlike David Hahn, Taylor made no attempt to hide anything; whenever the UPS man delivered something new, Taylor would explain it and show it off: an old GI compass, a bit of radiated graphite from a nuclear reactor, a gnomelike figurine made in Germany with a radioactive black glaze painted on the bottom. "It's uranium trioxide," he said, admiring it as he turned it over in his hands, "the same stuff they make nuclear fuel pellets out of."

When Tiffany and Kenneth questioned him about safety, Taylor spoke convincingly, if bafflingly, about time doses and distance intensities, of inverse-square laws and roentgen submultiples. He explained the different types of radiation and how they dictated the amount and kind of shielding necessary. For instance, alpha particles—the bound-up combination of two protons and two neutrons thrown off by uranium and its daughter elements—can easily be stopped by a sheet of paper or human skin, so they pose little external radiation threat. Higher-energy beta particles can travel through human tissue and cause serious damage, though they can be stopped by a few millimeters of aluminum. Gamma and x-rays carry the highest energy and are the most worrisome, since they can travel long distances and penetrate deep inside the body.

With his newfound knowledge, Taylor insisted, he could master the furtive energy that seeped from his rocks and metals and other collectibles. "I know what I'm doing," Taylor reassured his parents. "I'm the responsible radioactive Boy Scout."

• • •

But could an eleven-year-old, even one as science-savvy as Taylor, truly comprehend the consequences of a mistake? Acute radiation exposure — a devastating, one-time dose of ionizing radiation — is rare, but when it does happen, it can cause irreparable DNA damage and death. Chronic radiation syndrome — the result of long-term exposure to excessive radiation levels — is more common; it can lead to an increase in the likelihood of cancers and other radiation-induced diseases.

The health risks of nonacute exposures are difficult to gauge. For example, the 1985 Chernobyl disaster in present-day Ukraine led directly to 31 deaths (28 from acute radiation exposure) and 203 hospitalizations. But the number of eventual casualties caused by the radiation plume that spread across Europe will never be known. Estimates vary from four thousand additional cancer deaths above the normal rate (according to the World Health Organization) to more than sixteen thousand (per the International Agency for Research on Cancer).

Chronic exposures usually spur the body's natural DNA-repair mechanisms to compete — not always successfully — with the damage done by the radiation that passes through living tissue. The dangers and safeguards are better known now, but illnesses due to long-term dust and radiation exposure were common through much of the twentieth century. More than six hundred American miners, including many Navajo in New Mexico, suffered early demises after years of breathing uranium dust and radon gas. In eastern Germany's Ore Mountains, which supplied uranium for Soviet nukes, an estimated sixteen thousand miners were killed by the particles they breathed (though the health effects of coal mining are substantially more severe than uranium mining). Marie Curie's daughter Irène Joliot-Curie died of leukemia ten years after a vial of polonium broke open — but not before winning the family's third Nobel Prize (jointly, with her husband, Frédéric Joliot-Curie, in 1935) for discovering artificial radioactivity, the process by which elements are made radioactive by bombardment with charged particles.

The Curies never fully realized how dangerous their discoveries were. But there's evidence that, as early as 1917, owners and scientists at the United States Radium Corporation in Orange, New Jersey, understood the risks of radium exposure. They took precautions — for themselves — while assuring the low-paid Radium Girls who painted watch dials that the radium paint was harmless. Supervisors told the women to lick the brushes to keep their points sharp. For laughs, some women painted their fingernails and teeth with the glow-in-the-dark material. When they began to suffer bone fractures, anemia, and necrosis of the jaw, some demanded investigations and compensation. Company officials instructed the medical professionals they hired to attribute the ailments to other causes. They also claimed that those who'd joined a lawsuit had contracted syphilis, in an attempt to smear their reputations.

In terms of relative cancer risk, an absorbed dose of ten millirems (mrems) of ionizing radiation is equivalent to smoking 1.4 cigarettes, according to Idaho State University physicists. ("I think I've smoked a few cigarettes," Taylor would joke at one point, "but I follow the ALARP rule—'as low as reasonably possible.' When it comes to radiation hazards, I'm an ALARPist, not an alarmist.") For reference, a cross-country flight will give you a dose of about four mrems; a chest x-ray supplies between three and ten mrems; a set of dental x-rays produces a ten to forty mrem dose. Sleeping next to a friend for year (we are all radioactive) will give you a dose of about two mrems.

Natural background radiation is all around us, and there's little we can do about most of it. Humans have been immersed in radiation throughout our evolution; we've been showered from above by high-energy cosmic rays from the sun, supernovae, and pulsars. Particles from decaying elements in the Earth's crust bombard us from below, and radon gas (the second-leading cause of lung cancer, after smoking, in the U.S. and many countries) seeps into caves and basements. Bananas and potassium salt substitutes are measurably radioactive, as are lima beans and Brazil nuts. According to the U.S. Environmental Protection Agency (EPA), the average person is exposed to 360 mrems of background radiation each year from natural and manmade sources. That's far lower than it was in earlier stages of human evolution, when our species developed mechanisms to repair radiation-damaged DNA.

Manmade background radiation comes from nuclear test residue, medical devices, and coal-plant smokestacks, which spew fly ash containing thorium and uranium, dispersing far more radiation into the environment than nuclear power plants. Smoke detectors contain radioactive americium, and glow-in-the-dark emergency-exit signs, which are the most radioactive items the majority of people encounter in their daily lives, contain up to twenty-five curies of radioluminescent tritium, which emits low-energy beta particles.

That's not a problem—unless you take a sign apart and break open the vials of tritium while eating sunflower seeds, as one New Jersey teenager did. Tritium attaches to moisture-attracting tissue and can enter the bloodstream. State health officials, who moved quickly to decontaminate the boy's home in a costly cleanup, estimated that the sixteen-year-old absorbed about eighty mrems of radiation.

Through the mid-2000s, tritium exit signs were openly bought and sold on online auction sites such as eBay. Then collectors noticed auctions suddenly canceled, reportedly because of NRC intervention. In March of 2014, auctions of slightly radioactive trinitite (the glasslike residue left on the New Mexico desert floor after the Trinity nuclear detonation in 1945) were abolished from the site. Uranium ore, which is more radioactive than trinitite, is still available, presumably because the NRC doesn't place restrictions on small quantities of naturally occurring materials; only processed source material is subject to regulation.

Few would argue that trade in radioactive material (or other dangerous substances) should be an unregulated free-for-all. But the arbitrariness and clumsiness of enforcement calls into question the decision-making process and the efficacy of regulation efforts. On eBay, the long-standing prohibition of radioactive materials is selectively enforced and often creatively violated, as savvy sellers find ways to substitute flagged keywords with inventive euphemisms. "The problem," says one collector, "is that 'radioactive' describes anything and everything. Everything has some, whether it's a large or small emission. But if you put the word 'radioactive' in a listing you can guarantee it will be closed down."

Collectors like William Kolb, who coauthored *Living with Radiation: The First Hundred Years,* say the threat from radioactive collectibles is wildly overblown. "What hobbyists tend to pass along is not that dangerous. The stuff a terrorist could actually do real damage with would be something collectors wouldn't get near, like some of the old Third World medical machines."

In Goiânia, Brazil, in 1987, four people died and twenty were treated for radiation sickness after thieves stole and broke open an old radiotherapy source. In 2013, thieves in Mexico set off international nuclear-terrorism alerts when they hijacked a truck, unaware its cargo included a capsule containing three thousand curies of cobalt-60 from a decommissioned medical device. The hijackers got as far as opening its outer box before they heard a news report about the incident, freaked out, and dumped the contraband in a field outside Mexico City. Some of them raced to hospitals, fearing they'd been contaminated. But neither they nor the man who found the box and carried it home, hoping to sell it as scrap, got as far as opening the inner shielding. If anyone had opened it, even briefly, he would have been pierced by enough gamma rays to bring about a quick and almost certain death.

Other events can result in high levels of radiation that cause acute radiation syndrome, including escaped radioactive waste, a breached reactor, a solar flare during space travel, a criticality accident that initiates an uncontrolled fission reaction, and a nuclear explosion. A huge number of acute radiation cases occurred in the aftermath of the Hiroshima and Nagasaki bombings. Between 110,000 and 200,000 people died immediately from shock waves, intense heat, or radiation. Irradiated survivors staggered through the chaos, searching for relief from nausea, fever, hemorrhaging, and diarrhea. The most heavily dosed victims died within fourteen days from acute radiation syndrome. Those with lower doses survived longer; over the next five years an estimated 200,000 more died. Long-term survivors and their radiation-damaged offspring are known as *hibakusha*, or "explosion-affected people." Among them are numerous cases of thyroid and breast cancer, leukemia, and children born with mental retardation and extremely small heads, a condition known as microcephaly.

A few days after the Nagasaki bombing, one of the bomb's makers became a radiation victim himself in the world's first criticality accident. Working at night in the Los Alamos laboratory, scientist Harry Daghlian accidentally dropped a tungsten carbide brick onto a sphere of plutonium. A flash of blue momentarily lit up the room as air particles ionized and deadly neutrons streamed through the scientist. He desperately tried to knock off the brick, then began working to disassemble the tungsten carbide pile to halt the reaction and save his coworkers. But it was too late for Daghlian; he died twenty-five days later. The following year the same plutonium sphere, which scientists nicknamed "the demon core," went supercritical again during a botched experiment. Canadian physicist Louis Slotin shielded his coworkers and stopped the chain reaction by knocking the spheres apart, but he'd already absorbed nearly a thousand rads of gamma and neutron radiation; he died nine days later.

Available medical accounts of the casualties at Los Alamos, where every detail was classified, are sparse. But two recent radiation deaths — one accidental and one intentional — are well documented.

In 1999, two technicians were processing fuel at a facility northeast of Tokyo. While Masato Shinohara poured uranyl nitrate solution through a filter, Hiroshi Ouchi held the funnel. Though regulations required them to use a container specially shaped to prevent criticality, it had been set aside years earlier and replaced with an easier-to-handle bucket.

Suddenly, workers heard a loud crack and saw a burst of blue light. This light, often reported by observers of criticality accidents, is emitted when excited ionized atoms or molecules in the air fall back to unexcited states, producing an abundance of blue light. (This is also the reason airborne electrical sparks, including lightning, often appear blue.) A burst of neutrons penetrated the men's bodies and set off the radiation-alert siren. Ouchi and Shinohara ran into the next room, where Ouchi vomited and passed out.

When high-energy particles zap living tissue, they alter its atomic structure, stripping electrons away from atoms and molecules and breaking the chemical bonds that bind elements together. Once this chemical glue is gone, the cells essentially fall apart. Ouchi had regained consciousness by the time he arrived at the hospital; his face was slightly red and swollen, and his hand looked mildly sunburned. Doctors at first thought they'd be able to save him. But as authorities organized a suicide team to contain the out-of-control reaction, Ouchi's condition deteriorated. The neutron beams had converted sodium molecules inside his body into radioactive sodium-24 isotopes, which attacked the actively dividing cells in his blood, intestinal mucous membranes, and skin.

When doctors saw the first micrographs of the chromosomes in his bone-marrow cells, they gasped. What had once been long, orderly chains were now fragmented and scattered. With their DNA-repair mechanisms overpowered, Ouchi's mangled cells began, essentially, to commit suicide. His hair came out, and bodily fluids began streaming from his eyes, which sprang open every time the medical team tried to close them. As skin fell away from his body, fluids seeped out almost as fast as the team could replace them intravenously.

Ouchi's overwhelmed doctors reached out to experts around the world. But there were few precedents and no proven treatments for acute radiation poisoning. Mindful of a Japanese folk legend that a gift of a thousand origami cranes will grant the recipient a wish, his family stayed at his side folding paper birds as Ouchi became increasingly incoherent, calling for his mother and his home village. As his body lost its capacity to produce platelets, Ouchi dropped out of consciousness, his eyes finally closing. On the eighty-third day, after surviving longer than any criticality victim had, he finally passed away.

What happened to Ouchi was a preventable accident. Not so with the case of Alexander Litvinenko, a forty-three-year-old Russian dissident and former KGB agent whose tea was spiked with polonium during a November 2006 meeting with two other Russians in a bar in London's Grosvenor Square.

The highly ionizing particles emitted by polonium-210 can be blocked by skin or even a few inches of air. But once inside the body, alpha particles demolish the cells they encounter—whose contents, including the polonium, are absorbed by surrounding cells, creating an ever-expanding swath of devastation. Polonium-210, an almost pure alpha emitter, may well be the most toxic substance on earth; by weight, it's about two hundred and fifty thousand times more poisonous than hydrogen cyanide, the Nazis' favored poison. Litvinenko ingested just a speck of polonium, but it was many times more than the maximum safe body burden of seven-trillionths of a gram.

The Soviet Union apparently began using radiation as an assassination weapon in 1957, and today nearly all of the one hundred grams or so of polonium-210 made each year are produced at a repurposed weapons facility on the Volga River. Polonium-210 kills with devastating effectiveness, then departs the body quickly and leaves little trace —its half-life is only 138 days—which makes it a nearly perfect poison.

Litvinenko, an enemy of Russian president Vladimir Putin, vomited that evening, and his temperature dropped. Emergency-room doctors diagnosed a stomach infection and sent him home. Two days later he was back in the hospital, where baffled doctors administered antibiotics after tests found nothing conclusive.

As Litvinenko's condition deteriorated, his heart raced, pushing blood through radiation-swollen arteries and veins. Each beat delivered another blood-borne payload of radiation-emitting cells into his organs and eventually into his bone marrow, where corrupted cells produced ever more radioactive blood.

Before losing consciousness for the last time, Litvinenko opened his eyes briefly and told his wife he loved her, mumbling through the nubs of his lips, which he had, in his pain, chewed away. Then, mercifully, he died, twenty-two days after taking his deadly sip of tea. In that first instant of radiation exposure, Litvinenko's fate, like Ouchi's, had been inescapably sealed.

9

TRUST BUT VERIFY

ONE AFTERNOON, TIFFANY ducked her head into Taylor's garage laboratory to call him in for supper and saw her son in his yellow hazmat coveralls watching a pool of liquid as it spread from an overturned container across the concrete floor.

"Tay, it's time for supper."

"I'm going to have to clean this up first."

"That's not the stuff you said would kill us if it broke open," Tiffany asked, "is it?"

"I don't think so," Taylor said. "Not instantly."

Tiffany could have done with less of her son's drama. She had just lost her sister, her only sibling. And now, after years of remission, her mother's cancer had come back, so aggressively that Tiffany and Nell had flown to Chicago to see a specialist, who immediately started Nell on a regimen of chemotherapy.

Kenneth and Tiffany had other reasons to stress out that summer. Taylor had just graduated from the sheltering environment of St. James Day School, where he'd thrived. But the couple worried about how their quirky and willful son would fare at the public Texas Middle School, where it was unlikely that he'd find a teacher who'd indulge a ninety-minute monologue on nuclear physics or a principal willing to turn over computer monitors to a kid who wanted to disembowel them to make x-ray machines.

And then there was Joey, who seemed more and more overwhelmed by the hubbub that his brother created at home and wherever they went. While Taylor hammered away at radioactive rocks, played loud music, sang, and detonated homemade bombs, Joey increasingly withdrew into his own world. Tiffany and Kenneth had asked Joey if he wanted to go anyplace special during summer vacation that year, but he'd shrugged and said he was fine with going along with the rest of the family to New Mexico to prospect for uranium, a trip that Taylor had researched, suggested, and planned.

That July, Kenneth's daughter from his first marriage, Ashlee, then a college student, came to Texarkana for an extended stay. "The explosions in the backyard were getting to be a bit much," she says. "I could see everyone getting frustrated. They'd say something and Taylor would argue back, and his argument would be legitimate. He knows how to outthink you. I was telling my dad and Tiffany, 'You guys need to be parents. He's ruling the roost.'"

"What she didn't understand," Kenneth says, "is that we didn't have a choice. Taylor doesn't understand the meaning of *can't*."

"Looking back, I can see that," Ashlee concedes. "I mean, you can tell Taylor that the world doesn't revolve around him, but he doesn't really get that. He's not being selfish, it's just that there's so much going on in his head."

Tiffany and Kenneth had thrown themselves behind their sons' obsessions, enabling and encouraging them. But they were no longer dealing with the foot-stomping demands of a five-year-old who wanted a real crane for his birthday. Their eleven-year-old nuclear physicist might well be getting everything right, but if he *wasn't* . . . well, sometimes parents have to draw an unpopular line, especially when the consequences are so extreme. And Taylor had certainly gone far beyond the line most other reasonable parents would draw.

Why not just say no?

That is, of course, what David Hahn's parents did. And that's the point where the trajectories of Taylor Wilson and David Hahn began to diverge. When Hahn lost his parents' support for his scientific endeavors, he pressed on clandestinely, never again letting anyone into the secret world that was becoming an ever-bigger part of his identity. Once he took his hobby underground, there were no mentors to guide him, no one to check on his safety. That turned out to be a big problem.

Would Taylor have continued with his high-stakes collecting and experimenting in secret if his parents had banned his projects? When I put the question to him, he told me he was convinced that he would never have had to go there. He believed that if he made a rational case, he could talk his parents into anything. But I have little doubt that Taylor could have, and probably would have, gone underground with his scientific pursuits if he'd felt that was the only way to continue. Like

Hahn, he had a mix of will and curiosity that drove him to find his way around obstacles. And Taylor was ambitious and smart enough—at a much earlier age—to be at least as dangerous as Hahn had been.

Instead of instinctively doing what most parents would regard as common sense—keeping their child away from things that could kill him—Tiffany and Kenneth shifted to what was essentially a "trust but verify" policy.

"We thought, well, if he's going to get to where he needs to go, we've got to give him a little room to pursue it," Kenneth says. "The trick was to figure out a way to do that and keep everyone safe."

Though Kenneth was a natural problem solver, he had already run up against his own educational limits. "I was a business major," he says, "and I had to grit my teeth to get through the science requirements. I had a lot of faith in Taylor, and he could make compelling, data-supported arguments as to why it was safe. But he'd always talked a good game, and to be honest I was out of my league. I had to find people who had more technical knowledge, who could confirm that he wasn't just talking in circles."

And so he began reaching out to business, college, and Rotary Club connections.

Among those he called was a nuclear-pharmacist friend, David Boudreaux. Boudreaux, who owns Red River Pharmacy Services on the Texas side of the border, agreed to come over and check on Taylor's safety practices. The materials Taylor was working with required careful and specialized techniques to handle safely, Boudreaux said. He added that radiation worked in quick and complex ways. By the time Taylor learned from a mistake, it might be too late.

During sixth grade, Taylor had become consumed with the concept of nuclear transmutation, an idea that extended back to alchemists' dreams of transforming base metals into gold. The term *transmutation* dates to the Middle Ages, but it was first used in modern physics in 1901, when Frederick Soddy and Ernest Rutherford discovered that thorium was converting itself into radium through radioactive decay. Soddy yelled out, "Rutherford, this is transmutation!" and Rutherford retorted, "For Christ's sake, Soddy, don't call it transmutation. They'll have our heads off as alchemists."

Taylor's first attempts to induce artificial nuclear reactions followed the example of the experiments of Rutherford and Soddy, who shot an alpha particle into the nucleus of a light element in order to excite it, transform it, and release a free neutron. Taylor didn't bother trying americium-241 as a source ("At least that's one thing I learned from David Hahn," he says); instead, he skipped right to purified radium. Even then, he found that he couldn't produce enough neutrons to measurably transmute any of the materials he'd assembled.

Hahn, too, wanted to transmute elements. But shortly after he got his driver's license, at sixteen, he developed a new obsession; he wanted to build a breeder reactor that would generate more fissile material than it consumed. The resourceful Hahn managed to gather a hodgepodge of materials—radium salts, thorium ash, uranium powder, americium-241—and assemble them into a reactor that, he believed, could breed uranium-233 from thorium-232.

But Hahn's piecemeal scrounging left him far short of the critical mass of fissionable material necessary to create a chain reaction. (The inability to acquire sufficient quantities of these materials, rather than any technical roadblock, is what has thus far kept terrorists and madmen from making their own nuclear bombs.) Held together by duct tape, his so-called breeder reactor looked even less sophisticated than the model of a conventional reactor he had built for his merit badge using coat hangers, soda straws, and a juice can.

In the days after completing his reactor, Hahn noticed that each time he entered his shed laboratory, his Geiger counter clicked faster than it had the day before. Was the reactor actually working? The radiation levels were climbing; clearly, *something* was going on. One day, he took his Geiger counter outside and realized that radiation was detectable through the concrete foundation. Hahn began to worry that his reactor was heading toward an uncontrolled chain reaction of the type that had felled Ouchi, the Japanese technician, and the Los Alamos physicists.

More likely, the shed was getting increasingly contaminated with loose radioactive particles; he had far too much unshielded radioactive material in one place, and there was no question that there was a very real danger. When the detector began picking up radiation five doors away from his house, Hahn said, "I started to freak out." He quickly disassembled everything; he stored some of the materials in the house and shed and put the rest into a toolbox that he padlocked, sealed with duct tape, and placed in the trunk of his car.

A few hours later, Hahn was in jail, his car was impounded, and the police bomb squad had been called in, as had the state radiological experts. The discovery of radioactive materials in the trunk of Hahn's car had automatically triggered the Federal Radiological Emergency Response Plan. By the next day, the police and officials from two state agencies were joined by teams from four federal agencies.

"After that," said Hahn, "my life became a nightmare."

True to his Cajun heritage, David Boudreaux is easygoing and typically ready with a joke for just about any occasion. An avid duck hunter, he moves deliberately—a trait that serves him well in his business, which demands careful and accurate measuring. A nuclear pharmacist by training and an entrepreneur by nature, Boudreaux began building a chain of compounding pharmacies in and around Texarkana that now employs more than a hundred people. He grew his company by specializing in the sorts of things that most pharmacies don't have the expertise—or desire—to touch.

Red River Pharmacy Services supplies medical-imaging centers with radiopharmaceuticals. "We tag radioactive tracers to drugs, and those drugs carry them to the designated organ," Boudreaux says. "Then a gamma camera picks up that radioactivity coming out of the body and computer-enhances it into an image to show how an organ is functioning."

Boudreaux had come to visit Taylor at Kenneth's request. The two walked toward the garage, and Boudreaux asked Taylor about his most recent experiments. Taylor told him he'd been attempting transmutation, with disappointing results. He'd used a plastic-and-metal cylinder wrapped in paraffin to bombard beryllium, oxygen, and boron with alpha particles in an attempt to produce neutrons. He'd set up three detectors to measure different types of radiation.

Entering the garage, Boudreaux turned on one of the radiation survey meters he'd brought and asked Taylor what he was using as his alpha-radiation source. "When he said *radium,* I got what we Cajuns call a *frisson,*" Boudreaux says. "My industry has moved away from radium because it's so dangerous. But here was an eleven-year-old experimenting with the stuff that had killed some of the pioneers of nuclear physics."

Boudreaux managed to mask his reaction. "Tell me about your work plan for that experiment," Boudreaux said.

Taylor took Boudreaux through his process. "For radiation safety, there are three things to pay attention to," Taylor began. "Time, distance, and shielding." Taylor told Boudreaux that he organized his tasks to reduce the time of possible exposure. He had his radioactive sources well shielded, and he understood the inverse-square law (intensity is inversely proportional to the square of the distance from the source).

Boudreaux asked Taylor why such a detailed work plan was important. "Because," the boy replied, "it keeps you from needing to worry later whether you got it right."

"Right then, I knew I wasn't dealing with a normal eleven-year-old," Boudreaux says.

Taylor had straightened up his lab for Boudreaux's visit, arranging the usual chaotic jumble into half-organized piles. Boudreaux unpacked his own set of hypersensitive radiation detectors and made his way through Taylor's lab swabbing surfaces, testing, asking questions about protocols for storing and handling materials. Taylor showed his visitor the gloves he used to handle certain materials, the surgeon's mask he wore while working, the thermoluminescent dosimeter ring he placed on his finger during experiments.

"If anything, it was overkill for what he was dealing with at the time," says Boudreaux. "But it would turn out to be good practice since, as we know now, he had more ambitious plans."

They walked back to the house, where Tiffany and Kenneth were waiting in the kitchen. "No contamination," Boudreaux announced. He told them he was impressed by the way Taylor was handling himself and that they had nothing to worry about.

After that, Taylor and Boudreaux became good friends. Boudreaux would stop by from time to time to give Taylor spare equipment, and Tiffany or Kenneth would call him for advice on Christmas gifts for Taylor. "Whenever we got worried about something Taylor was doing," Tiffany says, "we'd ask David to come over again and check things out."

During dinner that night, Taylor and Boudreaux talked about how isotopes for diagnosing and treating cancer were made, shipped, and administered. "Some of it went over my head," Tiffany says, "but it turned out to be one of the rare conversations about nuclear stuff I could actually participate in, because of my mom's cancer and her radium implants."

In the 1960s, radium implants were still a fairly common approach to treating cervical cancer. Now, cesium-137 (which Taylor had used in his classroom demonstration) is much more common. Kenneth added that in the 1950s, his grandmother's thyroid goiters were treated with radioactive iodine-131 that, her doctor had told her, "came direct from Oak Ridge."

Taylor, who had recently read an article about the growing incidence of cancer in the developing world, asked how doctors in Africa got the isotopes they needed for their cancer patients.

"Unfortunately," Boudreaux said, "they usually can't." The expense of making isotopes, the special handling, and the high cost of just-in-time delivery systems have put diagnostic isotopes beyond the reach of much of the world's population. Even in North America, Boudreaux said, supply crises have developed because of the lack of backup capacity to produce some isotopes.

"I remember," Tiffany says, "that Taylor seemed upset that some people couldn't get the isotopes they needed to save their lives. He said it was sad. And then David said something that I didn't think much of then, but when I think of it now, I get a chill."

"Maybe," Boudreaux said, "Taylor will figure out a new way to do it and make all the stuff we're doing now obsolete."

10

EXTREME PARENTING

Boudreaux's willingness to take on the dual role of supportive uncle and expert mentor gave Taylor a new and nearby source of advice and camaraderie. For Tiffany and Kenneth, Boudreaux's involvement gave them the peace of mind they needed to allow Taylor to continue to collect and experiment. And it marked another milestone in what was an evolving parenting strategy that centered on finding opportunities to help their sons "grow and succeed and discover who they really are," as Tiffany puts it, "rather than controlling them and telling them who we think they should be."

On the spectrum of indulgence versus control, the Wilsons and Hahns were at opposite ends. Tiffany and Kenneth approached their sons' interests as opportunities to intensify their engagement with their children. Ken and Kathy viewed David's scientific pursuits as challenges to their authority that had to be thwarted, even though Ken and Kathy were, as engineers, more intellectually equipped than the Wilsons to support their young scientist and guide him toward his dreams. Instead, David's hands-on science interests became a wedge that widened the gulf between the generations.

According to dozens of gifted-education specialists I spoke with, Tiffany and Kenneth's approach was precisely on target in terms of talent development. The couple seemed to hit nearly all the right marks, at least in their children's early years, recognizing and engaging with Taylor's and Joey's gifts, staying involved and supportive without pushing them, letting them take intellectual risks, and connecting them with resources and mentors and experiences that allowed them to follow and extend their interests.

Psychologists and educators are quick to point out the difference between this sort of responsive parenting, in which a child's chosen interests are supported and encouraged, and "helicopter parenting," in which overcontrolling and overperfecting parents hover over their children's every move.

"So often what we see with these bright kids," says Ellen Winner, who chairs the Department of Psychology at Boston College, "is these very competitive parents steering their children toward things the parents have chosen." Far better to let children pilot their own helicopters and then get on board with the fuel—time, supplies, mentors, encouragement, and other resources—that can help children explore and extend their interests, whatever they may be.

What was evolving organically in the Wilson family was a remarkable climate of intellectual spoiling. And yet, if you ask Kenneth or Tiffany how they hit on the secret to raising smart kids, they'll tell you that they really don't know. Actually, Tiffany will first joke that the magic formula was "health food and Baby Einstein videos." Then she'll say that, in all seriousness, she and Kenneth didn't, at first, put a whole lot of thought into how they'd raise their children. They paid no attention to parenting books; they ignored the advice of other parents; they resisted the often reflexive inclination, which most parents have to some degree, to respond to their children in the ways their own parents had.

"Looking back, the way we parented was almost totally influenced by the kind of first child we had," Tiffany says. "Taylor was born and we took off on a crazy ride. A lot of the time we were just hanging on, winging it, following our instincts."

Although teachers, mentors, and others may influence how fully a talented child develops his or her potential, parents are usually the prime architects of a child's environment and are essential to bringing a young person's talents—whether common or uncommon—to fruition. "For things to go well depends more than anything on the actions of these crucial early catalysts," says David Henry Feldman, who directs the Developmental Science Group at Tufts University. "And of all the variables a gifted child faces, that's the hardest one to get right."

Feldman began studying profoundly gifted children and their cognitive development in the 1970s and spent ten years tracking six extraordinarily gifted children, whose stories he tells in his 1986 book *Nature's Gambit*. Though the body of knowledge about prodigiousness is still thin, Feldman's research convinced him that, in terms of supporting the development of a prodigy's talents, a parenting strategy like Tiffany and Kenneth's was ideal. "Taylor's parents may not have felt they knew what they were doing, but as luck would have it, their intuition was appropriate for the situation, and they had the wherewithal and the motivation. They were willing to take some risks and to spend time and energy enabling their son's very unusual pursuits."

But just as Taylor's talent for science and Joey's knack for math seemed to spring from nowhere, so did Tiffany's and Kenneth's responses. Where did their effective but counterintuitive reactions come from?

"I don't actually believe in reincarnation," says Feldman, "but I do believe that just as intellectual capacity is inherited to a certain extent, so too are parenting responses." Though Feldman stresses that he is speculating—very few prodigies have been intensively studied, and in any case his hunch would be almost impossible to test and prove—he believes that parents unknowingly bring a lot with them from previous generations. "If there was another science prodigy somewhere in the family history—and I'm willing to bet my thirty-year career that there was, in Taylor's case—then there were parents who reacted to that prodigy," Feldman says. "Generations later, when children exhibit a certain behavior, it triggers certain reactions in parents that are part of their own DNA." What happens next to that child depends to a great extent on what those responses are. "Some parents get it right, some just get it wrong. And some respond in ways that may have been appropriate in 1716 in Bologna but not here and now."

Were Tiffany and Kenneth drawing on some sort of transgenerational experience as they learned how to parent their gifted children? Kenneth believes there might be something to Feldman's idea. "At least, I can't think of another way to explain it. Looking back, I can see that our gut reactions to Taylor grew into a more conscious approach. But at first, honestly, we were flying by the seat of our pants a lot of the time."

What Tiffany and Kenneth did understand, almost from the beginning, was that they had two boys who were, as Tiffany puts it, "not your normal kids in most ways." It was clear that both Taylor and Joey were very smart, but Taylor asserted his intelligence with an extreme willfulness and obsession that was by turns confusing, entertaining, and exasperating. Kenneth in particular was baffled by his son, who was so profoundly different than him. Kenneth loves to watch and play the all-American games; Taylor has zero interest in team sports. Kenneth enjoys polishing off a plate of barbecued pork; Taylor, when he can be coaxed into eating, picks at vegetables and an occasional bit of organic free-range chicken breast. As Taylor grew older, he was drawn toward center-left politicians like Bill Clinton and Barack Obama, while Kenneth tended to be more conservative politically.

"I figured out a long time ago that Taylor wasn't interested in what I was interested in," Kenneth says. "We did expose him to everything normal kids do: T-ball, soccer, skis, Scouting, tennis. He's agile and he would have made a good athlete if he'd applied himself. But it was all too regimented for him. I tried to get him into golf, but he had no time for that either. He was all focus; he knew exactly what he wanted to do. And I just had to accept that and adapt and do things differently."

Most significant, Kenneth and Tiffany adapted by opening up opportunities that were outside the mainstream of what's available to most kids in southern Arkansas. Plenty of parents support their offspring's interests by buying things for them or dropping them off at the best schools or art centers that money can buy; far fewer put real time and effort into creating customized, hands-on opportunities that meaningfully expand their children's — and often their own — range of experiences.

Linda Brody, cofounder of the Diagnostic and Counseling Center at the Johns Hopkins University Center for Talented Youth, says that "the first trick is to give kids a lot of exposure to things in their younger years and then to notice what they're picking up on."

"Taylor and Joey weren't asking for baseball gloves or electronic gadgets like the other kids," Tiffany says. "So we introduced them to a lot of things to find out what they were interested in — in Taylor's case, he let us know in no uncertain terms — then we looked for ways to open doors that brought them in deeper."

It didn't hurt that they had a lot of community connections, such as Kenneth's friend who was willing to bring the crane over for Taylor's birthday or another family friend who arranged for Taylor to ride along in the helicopter that delivered Santa to a Christmas shopping-center event. When Taylor was researching nuclear reactors, Kenneth called an Arkansas senator who was then pushing for funding to decommission the Southwest Experimental Fast-Oxide Reactor, and the senator arranged for a tour of the shut-down facility, during which officials invited the eleven-year-old to climb inside the nuclear reactor's core.

Kenneth invited Taylor to a Rotary Club meeting to talk about the experience and about his interests in nuclear power and radioactivity. Kenneth, like teacher Angela Melde, had picked up on the learning-by-talking trait in Taylor, and he wanted to encourage it. As the date for the Rotary presentation approached, though, Kenneth began to get nervous. "I kept asking Taylor if he was going to practice, but he didn't prepare at all, he winged it. I didn't know what was going to happen, but he pulled it off; he brought down the house."

As their sons' interests expanded, the couple looked beyond Texarkana for ways to open up learning opportunities. When Joey expressed an interest in cooking, Tiffany found a class in southwestern cooking in Santa Fe. While Joey and Tiffany explored the possibilities of poblano peppers and green-chile salsa, Taylor and Kenneth drove to Los Alamos and visited the Bradbury Science Museum. Then the family reconnected to ski and visit the ancient adobes in Taos Pueblo.

"Take your kids places," say talent-development experts, who can now rely on an extensive and growing body of evidence that suggests that a lasting capacity for creativity is enhanced by early exposure to unusual and diverse situations. Such exposure inspires kids to connect the dots and recognize that "there are a lot of different ways of looking at different things," says Dean Keith Simonton, a psychologist at the University of California, Davis, who has written widely on genius and creativity. Early novel experiences, new psychological research suggests, play a substantial role in shaping the healthy development of brain systems that are important for effective learning and self-regulation, in childhood and beyond.

"We never had qualms about pulling the boys out of school for family trips that fed their interests," Kenneth says. "Even if they might go down a few points on a test, they'd learn something they otherwise wouldn't, something that might inform something else down the road."

Supporting Taylor's scientific adventures at home would prove to be a much bigger challenge both practically and emotionally. It's one thing to feed the talents of a kid who's into geology or horses. It's quite another to appropriately support a child who wants to experiment with the kinds of materials that keep the world's leaders awake at night.

People like Dodds, Boudreaux, and later mentors could double-check on Taylor's safety protocols and would eventually reach out to their own networks to help Taylor access the material and intellectual support he needed. As Taylor's nuclear ambitions grew, Tiffany and

Kenneth struggled even more to resist their urge to rein in their son. "It wasn't easy," Kenneth tells me, "but we realized that some kids come into this world with a special gift, and we couldn't keep that from him."

"Sometimes," Tiffany says, "the hardest part is to not stand in their way."

It was about to get a whole lot harder.

11

ACCELERATING TOWARD BIG SCIENCE

AFTER BOUDREAUX'S VISIT, Taylor felt he had a green light to get more ambitious with his collecting and his experiments, and his parents began to relax. Taylor stepped up his transmutation experiments, chasing the alchemist's ambition of transforming elements by irradiating stable isotopes and making them unstable. But none of the naturally radioactive sources in his expanding collection were able to give him the robust reactions he'd hoped for. Taylor was getting frustrated: "I realized that even if I collected lots more naturally radioactive stuff I might still not have sources powerful enough to create the isotopes I was interested in."

Ernest Rutherford came to the same realization as he used alpha particles from decaying radioactive materials to bombard atoms and investigate the nucleus. It was obvious to Rutherford and other physicists that their transmutation experiments were limited by the natural radioactive sources at their disposal. To move nuclear physics forward, they needed a way to crank up the energy of ion beams and accelerate the particles to higher speeds.

In 1929 Princeton professor Robert Van de Graaff designed a generator that could produce 80,000 volts of static electricity. This was the breakthrough physicists needed to take atomic science to the next level.

Cambridge researchers J. D. Cockcroft and E.T.S. Walton further multiplied the peak output of the first Van de Graaff generator design to power a device that could accelerate subatomic particles to unprecedented speeds and then slam them into target atoms, theorizing that the high-energy collisions in their atom smasher would create other subatomic particles and high-energy radiation such as x-rays.

In 1932, Cockcroft and Walton directed accelerated proton beams at a lithium target. The particles hit the lithium nucleus and transmuted the lithium into unstable beryllium, which then broke down into two helium nuclei, or alpha particles, releasing some seventeen million electron volts of energy. When the researchers detected alpha particles, they realized they had induced the first artificial nuclear reaction.

The result, which Cockcroft and Walton expressed in the equation ${}^{1}_{1}\mathrm{H} + {}^{7}_{3}\mathrm{Li} \rightarrow {}^{4}_{2}\mathrm{He} + {}^{4}_{2}\mathrm{He} + 17.2\ \mathrm{MeV}$, garnered the 1951 Nobel Prize in Physics and was the first verification of Einstein's equation $E = mc^2$. It also proved the value of accelerators, which would quickly become as important to the world of nuclear physics as telescopes had been to astronomy. The development of accelerators heralded the arrival of the era of Big Science, in which progress in high-energy physics and other scientific specialties would become increasingly dependent on large-scale projects using enormous machines in well-funded national laboratories.

Though Taylor wouldn't have the help of any big labs, he was intent on pushing his transmutation experiments forward. But like Rutherford and Cockcroft and Walton, he needed a radiation source that was more energetic than any of his naturally radioactive materials. "I realized," he says, "that if I was going to create anything useful, like medical isotopes, I was going to need to make my own radiation."

Again, Taylor's course of discovery was following the course of nuclear science itself. Searching for ways to take his research to the next level, he turned to a favorite source, the anthology of *Scientific American*'s Amateur Scientist columns. Two articles by C. L. Stong caught his eye. The first, a 1959 piece titled "How to Make an Electrostatic Machine to Accelerate Both Electrons and Protons," provided details of a machine constructed by chemical engineer F. B. Lee. Lee, Stong wrote, "designed an electrostatic accelerator suitable for amateur construction which is similar to the Cockcroft-Walton machine but has more than twice its power." The particle beam was "capable of cutting the time of chemical reactions, of inducing mutations in living organisms, of altering the physical properties of organic compounds and of producing scores of other interesting effects."

Each of the Amateur Scientist articles was headed with a graphic outlining the project's cost to build and levels of difficulty, utility, and danger. Stong assessed the particle accelerator's danger level at "Danger 3: Serious injury possible." In 1971, Stong featured another particle accelerator in an article titled "How to Build a Machine to Produce Low-Energy Protons and Deuterons." Whereas the first accelerator had a diffuse beam designed for the amateur chemist, this one, prototyped by

Larry Cress, could meet an amateur physicist's need for a more sharply focused beam that could bombard targets with protons and liberate gamma rays with energies substantially above 250 kilovolts.

Taylor was fired up; he just *had* to build his own accelerator. The best approach, he thought, would be to make a hybrid of the two and update it to create a powerful atom smasher. "Then," says Taylor, "I could produce nuclear reactions in light elements, and produce neutrons too."

In an open area near his bedroom, Taylor began tacking up schematic diagrams of the machines that he then supplemented with notes, calculations, and pictures of different components and hardware he could substitute to improve the design.

"Looks dangerous" was Tiffany's first reaction. Indeed, Stong had printed follow-up cautionary comments to his 1959 article, including one from an engineer who felt that the article hadn't adequately warned of the "far from negligible" hazards of x-rays. Stong agreed that "the article might well have pointed out . . . that one invites trouble by remaining near particle accelerators when they are in operation."

In Stong's 1971 article, he upped his assessment of the risk level to "Danger 4: POSSIBLY LETHAL!!" and included a warning that "the proton beam and the products of the nuclear reactions are hazardous. In addition to emitting gamma rays, the machine can generate x-rays of substantial intensity."

Taylor told his parents that he was convinced he could build and safely operate a hybrid of the two machines. But both Kenneth and Tiffany had reservations when Taylor pointed out that the Van de Graaff generator would produce between 250,000 and 500,000 volts. Was their eleven-year-old really ready to step into the high-voltage, atom-smashing world of Big Science?

Just after school let out for the summer, Taylor rode into work with Kenneth, something both he and Joey did frequently. Taylor loved visiting the plant, especially now that workers were disassembling the bottling line. The large scrap piles of machinery and electric controllers were full of parts he might be able to use for all sorts of things—including the particle accelerator he'd build if only his parents let him.

In the car, Taylor talked nonstop about the experiments he'd perform, the isotopes he'd create. But Kenneth leveled with Taylor: He and Tiffany wanted to support their son's scientific progress, but this project concerned them. They were worried about him working with hundreds of thousands of volts, and about the other hazards mentioned in the articles—what were they called?

"X-rays," Taylor said. "And gamma rays. And neutrons."

In 1993, psychology professor Barbara Kerr, now at the University of Kansas, began a fifteen-year study on the development of talent and creativity in gifted children. One finding in particular jumped out: "There are critical times when decisions are made that will either take a child down a path toward engagement with his or her intellectual interests, or disengagement," wrote Kerr, who also edited the 1,112-page *Encyclopedia of Giftedness, Creativity, and Talent.* The study found that these critical times vary by gender, subject interest, and physical, intellectual, and emotional development, and that each crossroads is rife with opportunities and risks. Subsequent research also made clear that the support (or lack of it) that a child receives at these make-or-break moments can have a tremendous effect on future outcomes, creating self-reinforcing and often permanent repercussions.

The first critical points are typically a child's initial interest in reading and the timing of kindergarten enrollment. In recent years, it has become fashionable to hold back children—boys, in particular—even when they're already reading and intellectually ready for school.

"It's well intended, but it does a bright child no favors," Kerr says. "Maybe they'll be tougher on the playground or excel in team sports, but intellectually, it holds them back. With reading, gifted kids are often discouraged by teachers who make them wait until the rest of the class catches up."

Another opportunity arises when a child develops a strong interest in a particular subject. If a parent notices and has the motivation and resources, he or she can provide opportunities to develop that interest and talent. If a parent doesn't see it, a good teacher might. But with only three states requiring general-education teachers to be trained in identifying and supporting gifted and talented students, the spark of interest often dies. "It's amazing how often parents and teachers fail to see the signs of prodigiousness, or to take action if they do," says Kerr.

The cutoff points for noticing and engaging exceptional talent vary widely between disciplines and individuals. Researchers Rena Subotnik and Paula Olszewski-Kubilius found that children who started chess or music at young ages typically achieved higher levels of mastery as adults, regardless of the quantity of cumulative practice. Even within particular fields, there are significant differences between subdisciplines. Wind instrumentalists and singers typically aren't physically developed enough to excel early, while string instrumentalists often show talent and need to start serious instruction at a young age, says Subotnik, "or it will go away."

In athletic pursuits, certain "early specialist sports"—gymnastics, figure skating, diving—can produce thirteen-year-old Olympic standouts, but star turns in team sports typically come later. Among intellectual pursuits, writing contrasts with mathematics and chess in that signs of literary mastery appear much later. "Writers need more human experience to understand and incorporate into their craft," says David Henry Feldman.

An early interest in science is one of the major predictors of eventual success in scientific careers, but the cutoff points for acquiring interest and skills in the sciences are less predictable than in other fields, and Kerr's research has shown that gender expectations can thwart the development of scientific talent. "When a child shows a passion for something that a parent fears isn't consistent with other children of their gender," she says, "some parents still fear that their kids will be bullied for being 'different.'" Instead of supporting a child's passions, a parent might push team sports on a boy who's more interested in theater or art, or shower princess dolls on a girl who likes science or athletics.

The impact of those parental choices are amplified in the lives of gifted children, Kerr says, "because the double messages are so strong in terms of achievement versus preferred sex role. I counsel parents to err on the side of achievement, but it's very difficult. Parents remember their own childhood experiences, and how severely kids punish other kids for not being obedient to gender roles."

A parent is more likely to notice and feed childhood interests if they're related to the parent's own career. Subotnik surveyed children at specialized science high schools and found that 60 percent had one or both parents in a scientific field. It's harder if you're a child with a talent in an area that your family doesn't care about. "In immigrant families especially, children sometimes get negative reinforcement if they don't want to be, say, a doctor or an engineer," says Subotnik.

Psychologist Scott Barry Kaufman, a researcher at the University of Pennsylvania, warns against closing the door too early on the "seemingly untalented" who are not early bloomers, and he uses his own experience as a compelling example. Diagnosed with an auditory processing disorder, he was sent to a school for children with learning disabilities and, he says, "fed a steady stream of low expectations." When he was fourteen, a new teacher noticed his boredom and frustration and asked why he was still in special education.

"For the first time in my life," Kaufman says, "my mind was suddenly brimming with possibility as I wondered: what am I actually capable of achieving?" He now has a master's in philosophy from Cambridge and a PhD from Yale and is the author of the book *Ungifted: Intelligence Redefined*. Kaufman readily admits that his story is a single anecdote and that he may be "just an outlier." But because a child's gifts may not be readily observable at any single moment in time, Kaufman says it's "an egregious error" to label and sort children as "gifted or ungifted" too early or to suggest limits on what a child can ultimately achieve.

The trick, says Kaufman, is to find a creative outlet that best fits a child's unique set of characteristics. "Once you find that domain, the learning process can proceed extremely rapidly, as the individual becomes inspired to excel." He recommends that parents stay on the lookout for activities that bring on the state of focused motivation that Hungarian psychologist Mihály Csíkszentmihályi famously termed *flow*—a single-minded immersion in an activity in which one becomes so motivated and involved that nothing else seems to matter.

In Taylor's case, his enthusiasm came so early—and was so focused, so intense, and often so vocal—that it was impossible to miss. Working in his laboratory, he'd regularly enter the flow state, becoming so absorbed in his experiments that he'd forget about time, food, and everything else. It was relatively easy at the outset for his parents to nurture their son's evolving obsessions, even when he first got into nukes. But what had started with their son's science-fair assemblage of dinner plates and lantern mantles had evolved into his desire to build an atom smasher with the power to throw out several varieties of DNA-mangling radiation. Increasingly, Tiffany and Kenneth seemed to be faced with a choice between supporting their son's passions and keeping him alive. At some point, they realized, they might have to draw the line.

For teenage rocketeer Homer Hickam's father, that point came after his son's errant rocket roared through town, narrowly missing several miners and slamming into the mine's headquarters. The boy's obsession with rocket science had been inspired by the Sputnik launch and

Wernher von Braun's response. Homer and his friends had captured the townspeople's imagination and moral support as their rockets grew bigger and went ever higher. But they had also raised hell with a litany of botched launches, busted fences, fires, and explosions. To Homer's father, the mine's foreman, his son's hobby represented an embarrassing and dangerous lack of discipline.

The inadvertent rocket attack on the headquarters convinced the elder Hickam that it was just too dangerous—for the miners and their families, for his son, for his job and reputation—to let the launches continue. That evening, he came home, gathered Homer's chemicals into a box, and took it and his son to a nearby creek.

"This is the end of it," he said, pouring everything out. "And this time I mean it. Collect stamps, catch frogs, keep bugs in a jar, do whatever you want. But no more rockets."

That Sunday in church, the reverend looked directly at Homer, who slid lower in his pew, and quoted from the Bible: "A foolish son is the calamity of his father. Cease, my son, to hear the instruction that causeth to err from the words of knowledge."

What Homer didn't realize yet was that the teachers and women's club had lobbied the reverend to do whatever he could to keep the town's young scientists in business. The preacher again quoted the Bible: "He that begetteth a wise child shall have joy of him. To have a child who longs to learn is the sweetest gift of all."

The reverend, on a roll, continued:

> Sons, obey your fathers. But fathers, help your sons to dream. If they are confused, counsel them. If they stray, search them out and bring them home . . . Fathers, I beseech you to seek out your straying sons and rescue them by keeping their dreams alive. These boys, and we all know I'm talking about our very own *rocket boys,* are dreaming great dreams. They should be helped, not stifled.

The sermon turned the momentum back toward the rocket boys. After church, the elder Hickam tossed his son the car keys and directed him to a slack dump outside of town. Tons of coal tailings had been bulldozed to create a square mile of flat black desert, perfect for launching and recovering rockets and far enough from town to be safe.

Cape Coalwood would serve as the proving ground for the club's increasingly ambitious science-based rocketry program. Their ingenious designs would go on to win the National Science Fair, and Homer Hickam, the boy rocket scientist, would go on to Rocket City (Huntsville, Alabama) to become a NASA aerospace engineer.

Kenneth and Taylor arrived at the Coke plant and walked through the cavernous space that was morphing, week by week, from a bottling factory to a warehouse. They ran into Tom Chesshir, the service manager who repaired the company's vending equipment. Chesshir, in his sixties at the time, was a big man, an outgoing ex-athlete and coach who, as Kenneth puts it, "never met a stranger." The nature of his job—half electrician and half mechanic—and his farming background made him a great all-around handyman.

When Taylor mentioned his would-be project, Chesshir remembered that he'd once seen a Van de Graaff generator in operation at a science fair in Dallas, where he'd taught earth sciences for five years. "A guy put his hands on it when it was running, and his hair stood up on end," Chesshir said. "Just about the strangest thing I ever did see."

"And it didn't kill him?" Kenneth asked.

Chesshir explained that even though a Van de Graaff machine can generate hundreds of thousands of volts, the amperage (the measure of electrical current) is too low to be hazardous.

"Current is what kills," Chesshir told Kenneth, "not voltage."

It was something that Kenneth had missed or that Taylor had neglected to mention in his hyperkinetic sales pitches to his father, whose head was spinning with all the talk of kilovolts and gamma rays and neutrons. Taylor unrolled his blueprints of the Van de Graaff generator for Chesshir and took him through the design—leaving rolled up, for now, the plans for the other half of the project, the accelerator that physics pioneers had used to smash subatomic particles and release some seventeen million electron volts.

Chesshir was one of Kenneth's most trusted employees. "What would you think," Kenneth asked him as they looked over Taylor's plans, "about helping Taylor build this thing this summer?"

Chesshir liked Taylor and enjoyed working with young people. He said he'd love to help. He and Kenneth made a deal that Chesshir would come in a day or two each week, as his other duties allowed, and work with Taylor in a corner of the plant, using as much scrap equipment as they could.

Taylor, ecstatic, ran over to a pile of machinery he'd collected from the bottling line and showed Chesshir and Kenneth a few of the salvaged O-rings, knobs, and other equipment he thought would be suitable for the project. He pulled out an instrument panel with switches on the front and an octopus tangle of wires tumbling out of the back.

"And this," Taylor said, holding it up, "will be our control panel!"

12

HEAVY WATER

TAYLOR, WHO HAULED IN PARTS and tacked his diagrams to a wall in the corner of the plant's workshop area, says his role was lead designer. Chesshir's role, which he'd discussed privately with Kenneth, was to help Taylor find parts and build the machine and, mostly, Chesshir remembers, "keep Taylor from blowing us all up."

Three-quarters of a century before Van de Graaff invented his machine, James Clerk Maxwell published his landmark equations on electromagnetism. Until then, scientists had wondered how an electric charge could exert forces on distant charges and objects, how such a force moved across empty space, and how fast it traveled. Maxwell's explanations made it clear that electricity and magnetism were best understood as fields, which fill surrounding space and indirectly affect other charges and objects within the field.

A conventional Van de Graaff machine creates an extreme electrostatic imbalance by using a sphere atop an insulating column that separates two electrodes that are charged by a motor-driven belt. The belt pulls electrons from one terminal and dumps them in the other, building up a strong electric field between the terminals.

Taylor was just getting comfortable using power tools, so Chesshir guided him as they cut large pieces of metal and wired everything together. Taylor had hoped to build his machine using mostly scrap materials from the bottling plant. But once they got started, he and Chesshir ended up making trips to the hardware store or ordering something online nearly every day. They picked up a pair of extra-large stainless-steel mixing bowls at Walmart that became the hemispheres for the globes after they cut the lips off and joined them with aluminum foil tape. The trickiest part was the belt, which had to be custom made from neoprene strips.

"Taylor was always go, go, go," Chesshir says, "never wanting to stop and take a break. It was fun watching someone that ambitious with that kind of attention span, which is unusual for a kid today. But he wanted every step of the project done that day, right then. I'd have to tell him, 'Look, we're in Nashville, Arkansas, five thousand population; we're just going to have to send away for this.' That wait for a delivery would just about kill him.

"When we finally got the generator going," says Chesshir, "that was a big day! Here was this kid I'd known since he was two, when he was too shy to talk, and suddenly when that thing started sparking, his eyes were lighting up and he was jumping for joy."

One way to estimate voltages is by measuring by how far a spark will jump through the air. A static charge of ten thousand to thirty thousand volts jumps a half an inch or so; Taylor's machine could throw a spark five or six inches. "We had a whole lot of fun testing it out by shocking each other," Taylor says. "You'd get close and you could feel your arm hair standing up. I'd have Tom or Joey put their hands on it, then I'd run up and shock them, and they'd do the same to me. Even though the charge is small, it's enough to make you jump and give a little yelp."

With the power source finished, they turned to building the particle accelerator; here, Chesshir reached the limits of his expertise. "When I'd try to get Taylor to explain it," Chesshir says, "he'd give a big dissertation about subatomic particles and neutrons and I'd say, Never mind, just tell me what we need to get."

Taylor had managed to track down Larry Cress, the source of the plans for the proton/deuteron accelerator featured in the 1971 Amateur Scientist column. After they talked on the phone, Cress sent Taylor a big packet of journal articles and design documents.

Taking cues from the trove of materials, Taylor modified the plans. Lacking a TIG (tungsten inert gas) welder to connect aluminum and other metal parts, Taylor and Chesshir built the accelerator column out of glass, copper, and brass and attached it to a collecting dome identical to the one atop the Van de Graaff generator.

The original plans specified an ignition coil from a Model T Ford to generate the arc for the ion source. In this case, Taylor was able to stay true to the plan. "My grandfather had his Model T parked in the warehouse, and I asked him if I could borrow the spark coil from it. It turned out to be perfect for the job. It was high voltage and compact, so I could easily fit it inside the small space inside the accelerator."

If all went according to design, the arc would strip the electrons off atoms, creating plasma, the ionized gas consisting of free electrons and positive ions. These ions would then accelerate down the tube and collide and produce gamma rays (if they used a hydrogen source) or neutrons (if they used deuterium).

As the project took shape, Chesshir began to trust Taylor's judgment more and more. "I was impressed by how he was thinking everything through, improving the original plans. But some of that stuff I was ordering online—well, let's just say Homeland Security's not doing their job. I sometimes worried I might get arrested, because no one would believe I was just the gopher and it was an eleven-year-old running the show."

To produce neutrons, Taylor needed deuterium gas to fuel the accelerator. Deuterium, a stable isotope of hydrogen, is not radioactive. "But my parents equated hydrogen with the *Hindenburg* disaster," he says, "and they weren't confident enough to buy me a canister of deuterium gas. So I had to figure out a way to make my own."

Taylor ordered some heavy water (also called deuterated water) from United Nuclear Scientific Supplies. Heavy water, made from oxygen bonded to one or more deuterium atoms, is used in Canadian nuclear reactors to moderate the unenriched uranium fuel. To extract deuterium from the heavy water, Taylor modified a Vacutainer blood-collection tube and turned it into an electrolysis device that could split the liquid into deuterium and oxygen. "It was a pretty sophisticated little rig," he says, "and it worked really well."

The finished accelerator looked great. "When I compare it to pictures of the first cyclotron that Ernest Lawrence built [for which he won the Nobel Prize in 1939], I can see that our construction techniques are similar—although I think mine looked even better. We had that nice control panel off the bottling line, with lots of cool-looking buttons and switches and relays," Taylor says. "We had it wired up so it would start the pumps and the ion sources."

But Taylor hadn't realized how challenging it would be to produce a workable vacuum. For his particle accelerator to function correctly, he'd need to pump out most of the air inside the accelerator column to create an almost empty space for his subatomic particles to travel in. If any gas or air molecules were left inside the tube, the accelerated particles would collide with them and lose energy. "Imagine a freeway in Los Angeles and you want to go a hundred miles an hour," Taylor explains. "If you try that at rush hour, you're going to hit other cars. But in the middle of the night, it's wide open and you can go fast."

To pump the air out of the tube, Taylor used a refrigerator compressor and wired it to run backward. Then he opened a valve to inject a small amount of his homemade deuterium gas. "I was so excited," Taylor says. "Me and Tom got the generator up above two hundred thousand volts, and with the Model T arc, we tried to get plasma going." But even though they used higher-tech fasteners than Lawrence had in the 1930s, they couldn't create a good enough vacuum to sustain a plasma field or clear enough air molecules to accelerate particles to any measurable degree. They tweaked the fasteners and tried all sorts of sealants — silicon rubber, epoxy, "and a few other things," says Taylor. "We were using techniques from the sixties and seventies, and we modernized them, but with our expertise and the materials we had, we could only go so far. Most of it worked. But not the big picture."

Taylor didn't know it then, but the setback was probably among the best things that could have happened to him in the long term. By taking on a challenge that was beyond him, he was, likely unconsciously, putting the value of learning above the value of not failing. That, says Stanford University psychologist Carol Dweck, is a hallmark of what she calls "growth mindset."

"When students have a growth mindset," says Dweck, "they understand that intelligence can be developed. Students focus on improvement instead of worrying about how smart they are and hungering for approval. They work hard to learn more and get smarter."

Years of research by Dweck and her colleagues have shown that students who learn with this mindset show greater motivation in school, get better grades, and have higher test scores. They are not discouraged by failure; in fact, they don't really see themselves as failing — they see themselves as learning. "Setbacks can actually make kids more motivated rather than less confident," says Dweck, whose assertion is backed up by a poll of 143 creativity researchers. The most important character foundation for creative achievement, said the majority of the researchers, is the kind of resilience and "fail-forward perseverance" that Taylor showed.

Taylor says he was "a little disappointed — but not all that disappointed, because I learned a lot, and we had fun. Considering we were in a small agricultural town I think we did a pretty good job. It was a failure, but it was a failure that added to my body of knowledge."

Instead of giving up, Taylor kept stretching himself in the direction he wanted to go. He analyzed his mistakes, reached out, and found new resources. Surfing the web, he came across Fusor.net, an information exchange for amateur high-energy physics scientists. Among those frequenting the site was Carl Willis, then a twenty-three-year-old PhD candidate living in Albuquerque. Taylor noticed that Willis had a helium-3 neutron detector for sale, and he struck up an online conversation about the detector and helium-3, which is the rarest, most expensive gas on the planet.

Taylor also told Willis that he was eleven years old and that he had tried to build a particle accelerator and wanted to take another stab at it. He asked Willis for some tips on collecting equipment and materials.

After a short conversation, Willis realized that Taylor was not just another geeky kid with a passing interest in nuclear science, someone whose intellectual depths he could plumb with a few technical questions. "I'd been contacted by other young people, but Taylor stood out. He had an impressive level of maturity and passion for the subject." Willis says he was struck by Taylor's knowledge, his focus, and his ability to synthesize information.

Willis told Taylor he'd like to help him pursue his interest in building high-energy atom smashers. But . . . the kid was eleven. "First," Willis told Taylor, "I need to talk to your dad."

13

BRIGHT AS THE SUN

TAYLOR DOESN'T REMEMBER exactly when he understood that his grandmother was dying.

"We tried to downplay what was happening," Tiffany says, "and my mom did her best to stay her upbeat self. She lost her hair but she still had mental energy, and she still had the kids over there all the time."

Nell was a good actor. As her health declined—and as the family sought out one treatment after another, conventional and alternative—she was, Taylor says, "as fun as ever, just maybe not quite as energetic." During her last few months, though, Nell began losing weight, losing energy, spending more time in bed. The inevitable became clear to Taylor, and he was increasingly upset about the prospect of losing his grandmother—who, Tiffany says, "had always been his biggest, unconditional supporter. Taylor adored her."

The cancer was eating Nell up inside, and it was eating Taylor up to see her withering away. "He tried to block it out," says Tiffany, "but he was pretty confused that summer." One day he'd be supermanic, running around and yelling and blowing things up and being a nuisance to Joey and everyone else. Then he'd spend the next two or three days alone in his laboratory. "It was painful for him," Tiffany says, "and he coped by going deeper into his science stuff."

What Taylor was getting deeper into now was uranium chemistry. One afternoon, Tiffany found Taylor in the garage crouched over a five-gallon bucket, using her gardening trowel to stir what looked like a thick yellowish-brown mud.

"Whatcha making there, Tay?" she asked.

"Yellowcake," Taylor said.

Yellowcake? Wasn't yellowcake what they used to build weapons of mass destruction? Tiffany remembered it from the run-up to the Iraq War, when (ultimately fictional) reports surfaced that Saddam Hussein's government was trying to get yellowcake from Niger to fuel his (ultimately mythical) nuclear weapons program.

Taylor tried to explain that yellowcake—partially processed uranium—was misunderstood. It wasn't a weapon of mass destruction, he told his mom. In fact, it was even less radioactive than the raw uranium rocks that they'd collected on the family's prospecting trip to the New Mexico desert.

"But Tay," Tiffany said, "why do you want to make yellowcake?"

"I don't know why making yellowcake was so exciting to me," Taylor would tell me later. "But I'd already gone from collecting radioactive household items on up to nuclear fuel pellets. And I was on my way, maybe, to making new atoms with an accelerator. Trying to refine a radioactive ore into an intermediate product seemed like filling in a gap in my experience."

Taylor had gotten the idea from Willis, who, with Kenneth's okay, had e-mailed Taylor some up-to-date academic papers on accelerators. After the two began communicating regularly, Taylor asked Willis for his yellowcake recipe. Willis sent it along and told Taylor that the process should be relatively easy for him since he already had some experience with uranium chemistry.

"What I didn't know then," says Taylor, "was that the yellowcake would actually come in really handy for some of the things I would do later."

Our drive from the Red Bluff Mine to the Wilsons' home in the hills of southwestern Reno takes about two hours. As we pull in, Kenneth clicks the garage-door remote control but parks the SUV in the driveway. Taylor jumps out and ducks under the ascending door and into the three-car garage. The darkness rises like a curtain, pushed up by a line of light that illuminates a cluttered assortment of Geiger counters, lead pigs, glove boxes, sealed ammunition cans, and other containers that hog up nearly every square inch of space on the floor, tables, and shelves. Against one wall is a large, upright gun safe festooned with a sign reading CAUTION: RADIOACTIVE MATERIALS AREA. PERSONAL MONITORING DEVICE REQUIRED. AUTHORIZED PERSONNEL ONLY.

As Kenneth carries in a box of the uranium ore from the mine, Taylor searches for a place to put it. Finally, he directs his dad to set it atop a box of ore from a previous outing. The addition of the other two boxes from the trunk creates an unsteady stack of rock and buckling cardboard; Taylor stabilizes it by leaning the top box against a chart of the nuclides tacked to the wall.

I ask him what he plans to do with his new haul. "Not sure," he says, laughing at the teetering boxes. "But you can never have enough uranium ore."

He scratches his head. "Actually, I got an idea. How about tomorrow we use some of it to make us a batch of yellowcake?"

When I come out to the garage the next morning, Taylor has already arranged a half-dozen buckets, beakers, and jugs in a half circle on the driveway. "I started making yellowcake that summer back when Grandma was dying," he tells me. "So I always have a lot of mixed emotions every time I do it."

We peer into the containers at various yellowish-brown liquids and sludges and powders, plus a few of the rocks we'd collected the day before. "I've got several batches in various stages," he says, "so I can take you through the whole process. It's actually pretty easy."

We start by smashing chunks of ore into smaller nuggets. Nothing fancy here; each of us whacks at them with a geologist's crack hammer, wearing protective goggles and dust masks. When most of it is down to the size of peas, we screen it through chicken wire, then through a smaller mesh, into a five-gallon bucket. "If I have tougher ore," Taylor says, "I'll use acid to break it down first."

Taylor pours a solution of sodium bicarbonate and sodium carbonate over the top of the pebbles and dust. "Baking soda and washing soda," he says, stirring it with Tiffany's gardening trowel. "That's all this stuff is! Then you come out a couple of times a day to stir the leaching buckets and re-suspend the solids."

Taylor moves over to two buckets of slurry that he started the previous week, now separated into uranyl carbonate solution and thick mud. He filters and collects the solution in a flask, checks the acidity with a pH testing strip, adds some hydrogen peroxide, and sets the flask atop a hot plate. The liquid has the appearance of slightly glowing urine.

"So, now we've taken what we want out of the rock by making it into a liquid," Taylor says. "This next step turns it back into a solid, which is how they export it in from uranium-producing countries." We bring it to a boil for a few minutes, then cut the flame. We'll leave it overnight to precipitate the insoluble uranyl peroxide, then we'll filter off the liquid and slowly heat the sediment to remove the moisture, watching it turn from mud to paste to yellow powder.

As Marie Curie discovered, and as Taylor told his mother, yellowcake is far less radioactive than both the original ore and the waste sludge that Taylor stores in plastic bottles. That's because most of the ore's radioactivity is produced not by the uranium itself but by its "daughters," decay products such as radium-226, radon-222, and polonium-218. Despite the intrigue that the word *yellowcake* arouses, the stuff itself isn't very dangerous.

It's what you do with it that matters.

"Remember Dubya's famous 'sixteen words'?" Taylor asks and then quotes from George W. Bush's 2003 State of the Union address: "'The British government has learned that Saddam Hussein recently sought significant quantities of uranium from Africa.' That was about yellowcake. It turned out to be a totally fabricated story, but even if he had some, there was nothing he could've done with it."

Taylor and Willis have refined uranium further, creating uranium tetrafluoride (known in the uranium refining industry as "green salt") and isolating it down to almost pure elemental uranium metal. "But lacking any mechanism for improving it to weapons grade," Taylor says, "that's as far as we, or anyone, could go."

Manufacturers of nuclear fuel or nuclear bombs transform the green salt to gaseous uranium hexafluoride and then use a centrifuge or some other device to enrich it and convert it into a ready-to-use solid fuel. Those processes require a tremendous amount of infrastructure and know-how.

"And so we went to war because the president said Iraq might be getting nuclear weapons and we couldn't afford to wait," Taylor says, surveying his sloppy science project. "But I've got a more advanced operation than Iraq ever had right here in my garage."

Taylor had told me over the phone that he has one of the largest private collections of radioactive materials on the planet.

"Taylor does exaggerate," Kenneth had warned me. But in the amateur nuclear physics community, Taylor's collection is renowned, as is Willis's. Now Taylor shows me his spy-agency radiation detectors, his water samples from Fukushima, his Soviet Geiger counters, his germicidal black light—"which I use sparingly 'cause it can cause cataracts and skin cancer."

He pulls out a box of thoriated tungsten welding electrodes, which he promptly spills all over the floor. "This is a hobby that introduces you to some weird stuff you wouldn't otherwise know about," he says as he crouches to pick up the rods. "Tungsten is within a few decimal points of the density of gold, so if you plated these with gold and sold them to someone who checked the density, it would work out. You could make billions before they'd figure it out. That's why I tell kids, 'Go into chemistry!' "

The upright gun safe was a recent Christmas present, and probably a very good idea. "I use it for my high-activity sources, the weapons and dirty-bomb stuff," Taylor says. "There are some very valuable, very radioactive things in here. Some very dangerous things. Some very proliferatable things. Proliferatable," he repeats, pausing to think. "Is that even a word?"

Taylor unlocks the safe and starts opening pigs; he pulls out samples of thallium-204, some thorium compounds, a small bit of cobalt. "Aha," he says, grabbing tweezers and removing a small black pellet. "Here's a plutonium fuel pellet from the Kerr-McGee plant in Oklahoma—the one where Karen Silkwood worked." Silkwood was a union activist who in 1974 testified to the Atomic Energy Commission about the lack of safety at the plant and then died in a mysterious car accident while on her way to meet *New York Times* investigative journalist David Burnham. She was found to have been contaminated by plutonium.

"Let's put this away quickly," Taylor says, dropping it back into a pig.

"There are a few things in here," he says, rummaging around in the safe, "that I don't even want to show you." He grabs a small vial containing a clear liquid. "Here's one thing," he says, taking it out of the safe, "that I really don't like to take out.

"Because if it were to drop and break open . . ."

I ask him what would happen. He doesn't answer.

"Let's just put it away," he says.

He's putting on a good show. I glance at the yellow-orange hazmat jumpsuit hanging from the rafter. Taylor has, since the age of eleven, worked with radioactive materials that are potentially lethal if mishandled. Taylor loves to be dramatic, but I can't help wondering out loud what would happen if he made a catastrophic mistake.

"That," he says, "is something that I always keep in the back of my mind. But I'm much more worried about chemical exposures. There's enough arsenic on a shelf in the physics lab at UNR"—the University of Nevada–Reno—"to kill half a million men. I don't fear radioactivity, because I understand it. And that gives me the power to protect myself."

Later that evening, I find Taylor in the garage. He's bent over a five-gallon bucket, stirring—and so focused that he doesn't notice me when I come in. When I say hello, he glances up and tells me that he came out to check on the yellowcake batches. Then he looks down and stirs some more, frowning slightly. A minute passes before he speaks again.

"When I first started doing this," he says, "boy, that was a tough summer."

He moves to the next bucket. "But as bad as it was with Grandma dying and all, that urine sure was something."

Urine?

Now Taylor looks sheepish. He knows this is weird. "After her PET scan she let me have a sample. I checked it with a Geiger counter and it was so hot from the diagnostic isotopes that I had to keep it in a lead pig.

"The other thing is . . ." He pauses, unsure whether to continue—but, being Taylor, he's unable to stop himself. "She had lung cancer, and she'd cough up little bits of tumor for me to dissect. I got some college microbiology and medical texts and some biotechnology books, and I figured out how to make a serum. I put in nutrients like what you find in blood serum, salts and proteins. Then I put the cancer cells in the medium and cultured them and got them to grow for a while."

He stops stirring and looks up.

"Some people might think that's gross, but I found it scientifically very interesting."

With the passing years it's become clear that Taylor the eleven-year-old wasn't using science just to block out the pain of his grandmother's impending death; he was using science as an act of defiance against it. Everyone was saying, toward the end, that there was nothing more anyone could do for Grandma. But there was Taylor in the laboratory he'd built in his grandmother's garage refusing to accept it, thinking, *There must be* something *I can do.*

What no one understood at first was that, as his grandmother was withering, Taylor was growing, moving beyond mere self-centeredness. The world that he saw revolving around him, he was starting to believe, was one that he could actually *change* and make better. As he held the clicking Geiger counter over the toxic urine sample, an idea began to take hold.

What if nukes weren't just about power and awe and magic?

Nuclear medicine had already saved far more people than nuclear bombs had killed. What if those lifesaving medical isotopes could reach even more people, faster? What if people like his grandmother could get earlier diagnoses? The challenge, as David Boudreaux had laid it out, is that isotopes for diagnosing cancer are extremely short-lived. They need to be, so they can get in and illuminate the targeted tumors and then decay away quickly, sparing healthy cells. Delivering them safely and on time requires expensive handling—including, usually, transport by private jet from the multimillion-dollar reactors and cyclotrons where they are made to the distant cities where they are administered to patients.

But what if those medical isotopes could be produced more cheaply, and closer to the patients? How many more people could they reach, and how much more quickly could they reach them? With an earlier diagnosis, how many more people like Taylor's grandmother could be saved?

Taylor had already thought about using the radioactive elements he was collecting to irradiate materials and transmute them into medical isotopes. Shortly after Irène Joliot-Curie and her husband, Frédéric, discovered that they could create short-lived radioisotopes this way (the discovery that won them their 1935 Nobel Prize), medical researchers began experimenting with treating cancer by injecting patients with irradiated silver nanoparticles (diagnostic medical isotopes would come later).

But even in the best case, Taylor would be limited by the amount of radioactive materials he could collect and by the risk of spills and contamination. "I saw that using the naturally radioactive stuff was hit or miss," he tells me as he stirs the third bucket of yellowcake-in-progress. "Even if I eventually collected every isotope, I needed a way to make things radioactive that I could control more precisely."

That summer, as Taylor stirred his grandmother's toxic urine sample, inspiration took hold. He peered into the swirling yellow center, and the answer shone up at him, bright as the sun. In fact, it was the sun—or, more precisely, the process that powers the sun: nuclear fusion. What if, instead of creating those isotopes in multimillion-dollar reactors and cyclotrons and then rushing them to patients, he could harvest the neutrons released by nuclear fusion reactions and use them to transmute materials into medical isotopes? What if he could build a small, tabletop nuclear fusion reactor that could produce those lifesaving isotopes as needed in every hospital in the world? There would be no need for huge production facilities, for private jets, for expensive logistics.

His brain was racing now, imagining possibilities that went beyond himself, beyond his grandmother, far beyond his garage laboratory and his hometown. If he could design a reactor small and cheap and safe enough, people everywhere, even in the middle of Africa, could get earlier diagnoses at a small fraction of the current cost.

But the process of building a miniature sun on Earth—accelerating particles at speeds and temperatures robust enough to fuse atoms—is extraordinarily complicated, something that government-sponsored research laboratories spend tens of billions of dollars on. At that point in Taylor's life, only a couple of dozen people had managed to build a working fusion reactor without the support of institutions or governments. Carl Willis was one of them. But Carl had two advanced degrees, access to a high-tech laboratory, and a nuclear engineer's salary to spend on precision equipment. How could an eleven-year-old kid in southern Arkansas ever hope to make his own star?

PART III

14

BRINGING THE STARS DOWN TO EARTH

NUCLEAR FUSION IS, for starters, the opposite of nuclear fission. Fission splits apart hefty, unstable atoms like uranium or plutonium to release their energy, giving us electricity, atomic bombs, and nuclear waste. Fusion combines the nuclei of two light, stable atoms into a single, heavier atom that is slightly less massive than the sum of its parts. That extra mass is released as energy. Fusion energy is extraordinarily abundant in our universe; it's what fills our world with heat and light, and it's what fuels life on Earth. The sun—our planet's distant power plant—and other stars are powered by fusion. In fact, much of what we know of in the visible universe originates from nuclear fusion. Most of the elements we're made of, the food we eat, and the Earth itself were born of fusion reactions.

Like all stable stars, the sun is a self-perpetuating thermonuclear fusion reactor fueled by hydrogen, the simplest and most abundant element. Each second, some five million tons of hydrogen atoms are drawn into the sun's high-pressure plasma core, where they collide so forcefully that their nuclei fuse, transforming single-proton hydrogen nuclei into double-proton helium nuclei.

In 1905, Albert Einstein proposed his famous $E = mc^2$ formula, which made it possible to understand the conversion process that powers the sun and other nuclear fusion reactions. Einstein's equation explained the fundamental relationship between mass and energy: that the energy content (E) of a body is equal to the mass (m) of the body times the speed of light (c) squared. In a fusion reaction, with each coupling of atomic nuclei, only a tiny amount of mass is lost, but the amount of energy released is immense—many times the amount of energy needed to bring the atoms together.

Flowing out of his theory of special relativity, Einstein's $E = mc^2$ insight spawned a whole new branch of science known as high-energy particle physics. Physicists who work in this field thrive on $E = mc^2$

conversions. Without a thorough understanding of the equation it would be impossible to build fusion reactors or other particle accelerators, impossible to understand the behavior of particles that are constantly colliding, releasing dollops of energy, and transmuting into newly formed particles. Once physicists had a basis for understanding these reactions, they began to imagine the possibilities of unleashing energy by splitting or combining atoms.

Fusion power has long been the Holy Grail of energy production, since it offers the possibility of abundant, clean electricity. Unlike fossil fuels or nuclear fission, fusion could give us energy without pollution, long-lasting radioactive waste, or the threat of a catastrophic release of deadly radiation. Fusion power plants would most likely be fueled with deuterium, a virtually inexhaustible hydrogen isotope found in seawater; and tritium, which is bred in reactors from lithium, a plentiful element. Two and a half pounds of these fuels can produce as much energy as eighteen million pounds of coal. Apart from energy, fusion produces helium, a useful inert gas that doesn't contribute to climate change. Although a fusion reactor's walls become radioactive after sustained neutron bombardment and need to be periodically replaced, this low-level waste decays to a safe level much more quickly—a hundred years versus tens of thousands of years—than the waste that collects in yellow drums outside fission power plants.

Fusion energy could conceivably be used to power far-roving spaceships. But to many of the thousands of scientists working to create fusion power on Earth, the quest to produce electricity by fusing atomic nuclei is no less than a race to enable humans to continue to survive on this planet. Climate scientists understand that continuing to burn fossil fuels at anywhere close to our current rate (about 80 percent of our energy is derived from fossil fuels) will raise atmospheric carbon dioxide (CO_2) levels to about five hundred parts per million by midcentury. The runaway growth in greenhouse-gas emissions has thus far swamped efforts to take meaningful action, but much of the world is waking up to the changes already happening to our climate. By 2100, the effects of these human-caused changes will be impossible to ignore as they lead to a series of devastating and possibly irreversible ecological impacts.

Unfortunately, no existing form of renewable energy is ready to serve as a prime-time replacement for fossil fuels—at least, not at a price most people are willing or able to pay. "The world needs a technology that can be switched on within a few decades, and preferably a lot sooner," says physicist Steven Cowley, director of the Culham Centre for Fusion Energy and CEO of the United Kingdom Atomic Energy Authority, "one that's compatible with current power grids, affordable, non-polluting, and impossible to use to make nuclear weapons."

Fusion energy does all that. Theoretically, it's the perfect energy source: safe and clean, fueled by virtually inexhaustible resources, and carbon-free. But there's a catch: Though nuclear fusion has powered the sun for 4.6 billion years, it has never produced usable energy on Earth, despite more than six decades of effort by the world's brightest scientists and engineers. The problem isn't whether fusion *can* work. Physics laboratories, even a few individuals, have successfully fused the nuclei of light atoms together, liberating their copious energy. But to actually produce energy on a useful scale, simply fusing atoms isn't enough. A viable fusion reactor would have to produce more energy than is poured into it.

That's the tricky part. Because hydrogen nuclei are positively charged, they repel each other and will fuse only if they collide with enough energy to overcome this repulsive electromagnetic force. Only when the nuclei get close enough to be drawn together by what scientists call the strong nuclear force—the extremely short-range force (less than 0.000000000000001 meters) that binds atomic nuclei—can fusion occur.

The sun overcomes electromagnetic repulsion with its massive gravity and extremely high pressure and core temperature (roughly 27 million degrees F, or 15 million degrees C). On Earth, fusion requires much higher energies and temperatures—at least 180 million degrees F, or 100 million degrees C.

Thus far, the methods that physicists have used to create environments conducive to fusing atoms have required more energy than the actual fusion reactions produce. Though one set of experiments reached so-called scientific break-even (generating as much energy as the fuel put into the system contained), a viable nuclear fusion power plant needs to go beyond both scientific break-even and engineering break-even; the energy produced and sent into the electric grid must exceed all the energy needed to create and sustain the reaction.

Einstein's formula inspired the physics community to explore the possibilities of bringing the power of the stars down to Earth. Progress and funding for experiments were scarce until the 1950s, when some American physicists who had worked on the hydrogen bomb began to shift their research toward controlled thermonuclear reactions. The most promising approach seemed to be magnetic confinement fusion (MCF), which uses a powerful magnetic field to confine and stabilize a superheated plasma field where atomic nuclei collide and fuse.

Plasma, which consists of free electrons and positive ions, is the only state of matter in which self-sustaining thermonuclear reactions can occur. Plasma shares similarities to gases, but in gases, electrons remain in their normal state, bound to the nuclei of atoms. In plasma, electrons are stripped away from the nuclei (which become positive ions) and move around freely.

Fusion research in the United States was at first classified, but when technological progress stalled, the U.S. and the USSR came to an agreement, signed in 1955, to open a worldwide exchange of fusion research and technology for peaceful uses. When scientists from the East and the West got together, the American physicists discovered that they'd gotten the better end of the deal; their Russian counterparts had developed a much more elegant arrangement of magnetic fields that produced denser, hotter plasma. The Russians called their doughnut-shaped design a tokamak, short for "toroidal chamber with an axial magnetic field."

By the 1980s, most major physics laboratories in the U.S. had built their own tokamaks and were experimenting with ways to stabilize the swirling, turbulent storms of plasma. Physicists theorized that a bigger chamber volume would increase a tokamak's capacity to stabilize the plasma and give the ions more time to collide and react. In 1985, Ronald Reagan and Mikhail Gorbachev agreed to collaborate on a massive tokamak, now called the International Thermonuclear Experimental Reactor, or ITER. Since then, the European Union, China, India, Japan, and South Korea have thrown their resources behind the experimental reactor, which is currently being built at the Cadarache Research Centre in Saint-Paul-lez-Durance, France.

The twenty-three-thousand-ton machine, under construction since 2007, won't be operational until 2019 at the earliest. Physicists expect that ITER's system of superconducting magnets—the most powerful ever built; they can exert a combined force of sixty meganewtons, enough to lift thirteen million pounds—will confine the plasma effectively enough to achieve fusion's long-sought breakthrough: a starlike self-sustaining reaction that produces more power than it consumes.

The experiment's goal is to generate five hundred megawatts of electricity, enough to power roughly five hundred thousand homes. But while most of ITER's physics problems have been worked out, huge engineering challenges remain. Foremost is the question of what the vacuum chamber's walls will be made of, as the material must be able to withstand constant bombardment by the neutrons released by the fusion reactions (unlike other particles, neutrons can't be contained by magnets, since they have no charge).

If the experiment succeeds, the next step will be a larger demonstration power plant. But ITER's project leaders don't really know exactly what will happen when they flip the switch on their massive magnetic bottle, the most complex machine — it will contain more than 10 million individual parts — mankind has ever built. Their often-frustrating, decades-long experiment may usher the world into a new era of safe, clean, and abundant energy — or it may be a twenty-billion-dollar fiasco.

Meanwhile, a fundamentally different approach to nuclear fusion, called inertial confinement fusion, or ICF, shows promise. ICF machines typically use lasers to flash-heat a deuterium/tritium fuel capsule. Though the capsule is only the size of a pinhead, it has the energy content of a full barrel of oil. To release that energy, a circle of lasers hit the capsule's outer layer with a high-energy pulse, compressing and heating the atoms at the center to the point that their nuclei fuse. Ignition, in the case of ICF, would occur when these reactions force the surrounding fuel into fusion and create a sustained chain reaction. Although inertial confinement reaction vessels are much smaller than tokamaks, the lasers and the systems needed to power them are enormous. The main technical challenge of ICF is that when the lasers fire, some of the fuel in the pellets can escape before a significant portion of the fuel has a chance to undergo fusion.

The fusion physics community has thrown much of its intellectual and financial weight behind ITER, which is currently eating up between one billion and two billion dollars per year. At the current pace of progress and funding, it will be at least thirty years before fusion energy production becomes a reality by ITER or any other technology. Given the potential payoffs and high stakes — filling the world's energy needs for millennia to come and saving the planet from environmental catastrophe — we need to ask whether we're willing to wait that long.

In 1961, John F. Kennedy declared that America would send astronauts to the moon and bring them home safely by the end of the decade.

It was a bold idea and a difficult challenge, but the nation rallied behind the ambitious Apollo program. Inspired by the president's energetic vision and motivated by the perceived Soviet threat, politicians gave von Braun and his NASA team the resources they needed to fast-track the space race.

"What if we approached fusion energy like the U.S. approached Apollo?" asks Ralf Kaiser, a fusion physicist with the International Atomic Energy Agency (IAEA). "Instead of scaling back ITER funding [as has been done several times], what if we instead put it on an ultra-fast track, with the entire world making an Apollo-like commitment?"

Between 1963 and 1972, the U.S. invested a little more than a hundred billion dollars (in today's dollars) on lunar programs. An Apollo-like budget (which would be underwritten by thirty-five countries rather than just one) in the neighborhood of ten billion dollars a year would cost just one-seventh of 1 percent of the seven trillion dollars the world spends each year on energy. With that kind of funding, top fusion physicists believe they could deliver fusion power within a decade. Betting that big on fusion would take a lot of political will and imagination —which was exactly what made Apollo possible.

Of course, the need for a safe, cheap, nonpolluting energy source is much more pressing than the need for a man to walk on the moon. And getting fusion to work will be much harder. "Getting to the moon is almost trivial in comparison to nuclear fusion energy," says plasma physicist Ron Phaneuf of the University of Nevada–Reno. "Fusion is by far the most significant scientific and technical challenge mankind has ever attempted."

All of which made nuclear fusion such a compelling challenge for Taylor, who already, at the age of eleven, seemed incapable of attempting anything on what most people would consider a reasonable scale.

"Someone saying it can't be done, or it's extremely hard to do, just makes me want to do it," Taylor says. "I just don't accept that I can't. I really do think that someday we'll have fusion power and that I can be part of the breakthroughs that make it happen."

But first, he wanted to use nuclear fusion as a means to generate neutrons. At eleven years old, Taylor was just beginning to see the connections among the disciplines he'd studied, and he was captivated by the idea that he might find answers to some very big questions hidden in those connections.

E. Paul Torrance, the noted creativity researcher, wrote elegantly about "the dreadful importance of falling in love with 'something'—a dream, an image of the future":

> Positive images of the future are a powerful and magnetic force. These images of the future draw us on and energize us, giving us the courage and will to take important initiatives and move forward to new solutions and achievements. To dream and to plan, to be curious about the future and to wonder how much it can be influenced by our efforts, are important aspects of our being human. In fact, life's most energizing and exciting moments occur in those split seconds when our strugglings and searchings are suddenly transformed into the dazzling aura of the profoundly new, an image of the future.

Torrance's mid- and late-twentieth-century take on the importance of one's image of his or her own future has been affirmed by modern psychologists who have researched motivation, creativity, and achievement; we now know that our beliefs about our abilities and potential drive our behaviors and predict our successes even more accurately than past performance does. That sort of future-oriented inspiration, researchers say, is an oft-overlooked factor that can affect one's motivation and capacity to work persistently toward long-term goals and help one prevail over obstacles and setbacks. There's also evidence that these motivational/inspirational factors can boost both cognitive efficiency and overall productivity.

From that point on, Taylor went forward not just out of curiosity but out of a genuine love for his subject and a desire to create something that could change the world and make it better. His ambition was not, for an eleven-year-old, particularly realistic. But the biographies and memoirs of many world-changers make it clear that the source of the sustained creative energy that fueled their breakthrough achievements was an early obsession with something that stayed with them for their entire lives.

Anthony Fauci, the renowned immunologist whose teams have achieved several breakthroughs in the treatment of HIV/AIDS, maintains that a lack of experience can at times actually be a benefit, since younger people, who are natural outsiders, often have less constricted, more creative views of the world.

"The trick is to work with young scientists and put them on projects that they don't know can't be done," says Fauci, who directs the

National Institute of Allergy and Infectious Diseases. "When you have experienced and inexperienced people working in the lab together, you can balance the skepticism that accomplished scientists often have. They'll say, 'You'll never be able to do that, why would you waste your time?' But young people are more open to taking a chance and so you often get a very important observation because someone went down a road a more experienced person wouldn't have."

Marie Curie was convinced of a causal relationship between the innate curiosity of childhood and great feats of discovery. And Einstein, well known for his childlike nature, insisted that "imagination is more important than knowledge. For knowledge is limited to all we now know and understand, while imagination embraces the entire world, and all there ever will be to know and understand."

"I think kids are able to sometimes do better science than adults," Taylor would tell me when he was on the verge of becoming an adult. "Because kids haven't been exposed to the bureaucracy of professional science, they're a lot more open to trying things."

If nuclear fusion is, as many physicists believe, still waiting for its big idea, then the notion that it could come from a young, forward-thinking physicist isn't as far-fetched as it might at first seem. Dean Simonton, who has studied the advantages of youth in the creative process, found that physics and poetry, more than other fields, tend to produce early bursts of productive inspiration. Although breakthroughs most often come from midcareer scientists, physicists frequently make their *first* important discoveries before the age of thirty.

Both Newton and Einstein experienced a "miracle year" in their midtwenties. Newton was twenty-four when an apple falling from a tree led him to hit upon the law of universal gravitation; that year, he also did important work in calculus, optics, and the laws of motion. Einstein's annus mirabilis, which still evokes awe in the world of science, came when he was twenty-six. That year (1905), he overturned Newton's idea that space and time are absolute with his special theory of relativity. He also developed the quantum theory of light, proved that atoms exist, explained Brownian motion, and described the relationship between energy and matter with $E = mc^2$.

Carl Willis, the nuclear engineer from Albuquerque, detected in Taylor those sparks of curiosity and passionate ambition that, coupled with an early command of nuclear physics, gave the boy "an ability to make logical and useful connections that would have been exceptional at any age." To get Taylor started, Willis sent him some specifications on a type of reactor known as a Farnsworth-Hirsch fusion reactor. Invented by Philo T. Farnsworth (who also invented the television) and improved by Robert Hirsch (who directed the U.S. fusion energy program during the 1970s), the fusor consists of a vacuum chamber surrounding one or two spherical inner grids that cradle a plasma field. With a just-right combination of fuel, vacuum pressure, and voltage, atoms ionize. As they become electrically charged, they accelerate toward the grid(s). When they pass through the grid(s) and converge in the central plasma field, some collide and fuse, releasing energy in the form of neutrons.

But Willis made it clear that getting a significant number of ions close enough to fuse and produce measurable quantities of neutrons would be a tremendous challenge—especially for someone with limited resources in a town lacking any technical infrastructure to speak of.

"But the thing about Carl," Taylor says, "was that he never said, 'You'll never be able to do it, you're just a kid and even if you weren't it would be too complicated and expensive.'

"He just said, 'It won't be easy.'"

A few days after they talked, Taylor received a parcel in the mail. It was a gift from Willis: the helium-3 neutron detector being sold on Fusor.net that they'd discussed during their first communication.

"That was a huge boost," Taylor says, "because it told me Carl believed there was a chance I'd actually be able to build a reactor that could achieve fusion and produce neutrons."

15

ROOTS OF PRODIGIOUSNESS

ONE SUMMER EVENING as a storm approached, Tiffany looked out the kitchen window and saw Taylor darting around the yard placing neutron detectors in trees and on the roof of the two-story backyard shed that the family called the Little House.

"I'm looking for neutrons that might-could be generated by lightning," he shouted in response to her query. He added, without breaking stride, that there was some evidence that lightning produced neutrons, but it hadn't been proven; it was an open question in physics.

"Maybe," he yelled, "I can be the one who proves it!"

"Better get yourself inside pretty quick before you get whacked by some of those neutrons you're hunting for," Tiffany hollered back.

Psychologists and educators say Taylor's sort of intensity is almost always apparent in gifted and talented children. "I call it the rage to master," says psychologist Ellen Winner. "I hear it from parents all the time; they say there's nothing that can keep their kids from what they want to be good at."

Definitions of giftedness vary across the academic and geographic world, but researchers agree that the common denominators include extreme curiosity and an intense urge to discover. "Truly gifted kids are almost always autodidacts, motivated from within," says Susan G. Assouline, who directs the University of Iowa's Belin-Blank Center, a leading research center focusing on gifted education. "Sometimes there's something holding them back from expressing it, such as boredom or autism," she says. (Recent research suggests a genetic connection between autism and prodigiousness.) "But on the inside, these kids want to be engaged; they want opportunities to discover something new."

The U.S. Department of Education defines the academically gifted as "students, children, or youth who give evidence of high achievement capability in areas such as intellectual, creative, artistic, or leadership capacity, or in specific academic fields, and who need services and activities not ordinarily provided by the school in order to fully develop those capabilities."

The UK Department for Children, Schools, and Families has a less verbose definition of gifted and talented students: "children and young people with one or more abilities developed to a level significantly ahead of their year group (or with the potential to develop those abilities)."

The National Association for Gifted Children (NAGC) estimates that there are three to five million academically gifted children in kindergarten through twelfth grade in the U.S. — roughly 6 to 10 percent of the student population. In Britain, the estimate is 5 to 10 percent; Turkey and India say about 3 percent of their students are gifted. But the NAGC admits that its number is little more than an educated guess; no one really knows how many children are gifted or whether the proportion of them in the student body is growing or shrinking. "To know for sure, you'd have to precisely define criteria for giftedness, then do a big epidemiological-type study canvassing a whole country," says Winner. "Gifted kids do seem more visible now, but it may be that because of advances in technology and communication more are showing up who wouldn't have been noticed before."

That's probably true. In the pre-Internet era, it's not very likely we'd have heard about someone like Akrit Jaswal, who performed surgery on a burn victim's hands at age seven and was admitted to an Indian medical university at twelve; or William Kamkwamba, who built a wind turbine to power his impoverished Malawian village at age fifteen; or Gregory Smith, an American who finished elementary school in one year and high school in two years. After graduating from college with honors at thirteen, Gregory is pursuing a PhD in mathematics and traveling to developing nations to promote peace and children's rights. He was nominated at twelve years old — and three times since — for the Nobel Peace Prize.

The community of gifted children includes a much smaller subset of kids often described as prodigies, or the profoundly gifted. These are the scary-smart kids whose talents and early achievements are off the charts. Among the handful of academic researchers who have studied them extensively, no one has devoted more time and energy than David Henry Feldman, the Tufts University child psychologist. Feldman defines a prodigy as "a pre-adolescent who is performing at the level of a highly trained adult in a very demanding field of endeavor." In contrast to an especially smart kid of great general ability, the prodigy has a distinct form of giftedness that's far more advanced and focused on a single interest.

These are the children who devour books (often nonfiction) before entering kindergarten; who teach themselves algebra or musical notation as toddlers; who take our breath away with their piano prowess, their devastatingly efficient chess moves, or their visionary artwork. With prodigies, the rage to master is extreme. They are attracted to a subject early and learn rapidly, approaching it with unshakable concentration.

Feldman became intrigued by profoundly gifted children and their cognitive development in the 1970s, and over ten years he closely tracked six extraordinarily gifted children. He told their stories in his 1986 book *Nature's Gambit.* The book's content is both surprising (among other findings, he noted that prodigies' IQs vary widely depending on specialty) and disconcerting, since prodigiousness, Feldman discovered, often does not lead to happy adulthoods. Feldman's sensitivity and commitment to his subjects are impressive, as are his observations of the characteristics that define and link these children. But he stresses that the body of knowledge about extreme giftedness is thin. Prodigies, by definition, are hard to gather in large numbers, and they are often too busy to participate in studies. "We're dealing with very tiny samples, maybe fifty children who have been intensely studied," Feldman says. "We are at the very beginning of understanding them."

The past decade has brought renewed academic interest in prodigies and gifted children. Researchers see the gifted child (along with other outliers, such as savants, autistic children, and very high-IQ cases) as among the more striking manifestations of human potential. Understanding their intellectual development is an important key to deciphering the complexities of intellectual development across the entire spectrum of children.

One of the primary questions puzzling researchers like Feldman is exactly where and how intelligence is generated in the brain. But the neuroanatomical basis of prodigiousness still generates more questions than answers. For instance, how do neurological reward systems differ between high achievers and other children? How and why are some brains seemingly wired to love novelty? Why do some people have such a driving need to ask questions and achieve mastery? What neural systems underpin the ability to outperform the rest of us? And how do external factors affect brain development?

"Unfortunately, we don't yet have the ability to answer, or even properly ask, what's going on inside a prodigy's brain that creates those conditions," says neuroscientist Francisco Xavier Castellanos, director of research at New York University's Child Study Center. In the near future, neuroimaging will likely boost our understanding of the brain's development and help us predict future performance and deficits, health-related behaviors, and responses to pharmacological or behavioral treatments. "But the tools we have now to observe the brain are a bit like an early-generation microscope," says Castellanos, who has spent much of his career scanning and analyzing the brains of children and adolescents. "We can tell that the brain is very busy, but we have profound ignorance about its workings."

Neuroscientists have observed a correlation between brain size and IQ—but it accounts for only about 10 percent of variations. Research also suggests that adult intelligence can be predicted during childhood by the rate of change in the thickness of a child's cerebral cortex, the gray matter that makes up the brain's outer layer. Other evidence has emerged that the efficiency with which information travels through the structures of the brain—in particular between the parietal and frontal lobes—may be a significant enabling factor for intelligence.

General intelligence is typically gauged by IQ score, which remains a popular scale in large part because it's a reasonably comparable measure that's been applied over a long term to large numbers of people. "IQ is like democracy; it's better than the rest but it has acknowledged flaws," says Castellanos. "We can't deny that IQ data are among the best measures in finding differences between people in general, but reliable ways."

And yet, IQ doesn't go very far in explaining the way smart people's brains work or in providing useful tools for mapping the path from raw cognitive ability to remarkable achievement. "[IQ is] unlikely to predict a kid who has abilities like Taylor's," says Castellanos. "There'd be lots of kids with IQs as high as his who wouldn't have gone this far, this fast."

Joanne Ruthsatz, a psychologist at Ohio State University, confirmed Feldman's observations that IQ requirements for mastery vary widely across specialties. In a 2012 paper published in the journal *Intelligence*, she reported that a typical IQ for an art prodigy is around 100. Early chess and music masters usually have higher IQs, as do math whizzes, who average in the 140 range.

A high IQ will not take you far in the art world (in fact, it appears to be a handicap for those with artistic gifts), but it's almost essential for a high-achieving physicist. Then again, there are exceptions. The iconoclastic physicist Richard Feynman had an IQ of 124, high but not spectacular, yet in his midtwenties he mastered the astonishingly complex equations that led to his Nobel Prize–winning theory of quantum electrodynamics (QED), and he did it with what biographer James Gleick called "frightening ease."

"Winning a Nobel Prize is no big deal," Feynman reportedly told his wife after accepting the prize, "but winning it with an IQ of 124 is really something."

The search for a specific cognitive ability that underlies all forms of prodigiousness has pointed most distinctly toward working memory, a key cognitive function that allows us to hold information in our minds in a highly active state. According to several recent studies, a high working-memory capacity is a constant that appears to predict and positively affect performance and intelligence across all domains. Ruthsatz has administered standardized IQ tests to prodigies and found that, although IQs ranged from 108 to 147—just above average to above the conventional cutoff for genius—each prodigy was at or above the 99th percentile for working memory.

So how does working memory work?

"It's been argued that the brain is limitless as to how much information can be stored in a lifespan," says Castellanos, "and no one has proven otherwise. What's profoundly limited is how many things we can hold in our mind at one time, available for processing. The magic number for humans seems to be seven, plus or minus two." To get to a novel thought, a person has to be able to maintain several things up in the air, ready to manipulate.

It's relatively easy to test for working-memory capacity. One common way to measure it is by seeing how many items someone can remember simultaneously for a short period. If you're given two ten-digit phone numbers and can repeat them back, you have a higher working-memory capacity than someone who can recall only seven digits of one phone number. Working memory lets you multiply large numbers in your head or remember the names of two new acquaintances while being introduced to a third. It's crucial to academic, professional, and social success, and it typically increases steadily through childhood and adolescence, peaking between the ages of eighteen and twenty-five. It begins to decline in our thirties, but it can be fortified and preserved through training or pursuits that demand its use. It is this premise that has driven the success of brain-training programs and apps that promise to make our minds more supple and spry—although there is absolutely no solid scientific evidence to back up these promises.

Surprisingly, humans and monkeys have identical working-memory capacities. But modern humans have learned, in a profoundly useful adaptation, to hack the limits of working memory through a process called chunking, in which information is analyzed and compressed into composite nuggets that are more memorable and easier to process. For example, we chunk when we hear *212* not as three discrete elements but as the telephone area code for Manhattan. This allows our memory systems to be more efficient and effective at grasping possibilities and extracting useful structure from raw data.

High achievers are able to chunk and superchunk in the domains they're focused on. "We all know someone," says Castellanos, "who amazes us at how fast they can think. You observe them and you say, 'Holy shit, you got there faster than I did. Now, if I had X percent more time I might have been able to get there too.' I think of intelligence as how much can you get done in the span of time available—whether it's a lifespan or the moment in which we're thinking."

One of the longest and most frequently debated aspects of high achievers is the degree to which they are born or made. Is a person's intellectual destiny dependent on inherited mental hardware? Or can someone starting at subprodigy levels reach genius-level acumen in a chosen discipline through ambition and long, hard work?

Much of the research on talent development has been driven by the urge to answer these questions and predict who will rise to the top. An extensive and growing body of evidence confirms that we can identify future innovators by the time they are teenagers. The same research has confirmed that those whose abilities are identified early and who are given support to develop their talents are the ones most likely to grow into the creative, high-achieving adults who transform society, advance knowledge, and reinvent modern culture.

"Exceptional youthful ability really does correlate with exceptional adult achievement," says Jonathan Wai, a research scientist at the Duke University Talent Identification Program and the author of *Psychology Today*'s Finding the Next Einstein column. "It really is possible to identify the kids who are likely to become future innovators." Wai has investigated different populations of intellectual outliers to better understand the talent-development process and the degree to which brainiacs are likely to become billionaires.

In 1969, Johns Hopkins University's Julian Stanley began administering the SAT college-admissions test to children under thirteen. Stanley also initiated the Study of Mathematically Precocious Youth (SMPY), which since 1971 has tracked early takers of the SAT who score near the top on the math or verbal sections of the test (some two hundred thousand students participate annually in the searches by taking the SAT in the seventh instead of the usual eleventh grade). Following the education and career paths of what has become a cohort of more than five thousand, SMPY has accumulated forty-five years of data that provide much of what we now know about early aptitude and subsequent trajectories of some of the nation's smartest children.

A high percentage of the children who were identified as extremely intelligent went on to become high-achieving adults. The top 1 to .01 percent (those whose IQs range from 137 to 160) typically joined the upper echelons of the STEM professions or became high-achieving humanities professors, powerful politicians, or successful journalists or novelists. The average MD or PhD has an IQ of around 125, while individuals with IQs above 160 often do brilliant work in mathematics or physics, where success is even more dependent on raw mental processing power.

The takeaway, it would seem, is that innate talent and high general intelligence — which are to a great extent heritable, like titles of nobility — are a first-class, high-speed ticket to advanced achievement.

Not necessarily. For one thing, cognitive abilities don't appear fully formed at birth; they develop over time through a complex interplay between nature and nurture. (Research has attributed genetic influence on human intelligence to between 30 and 80 percent of its total variance.) Among the most important discoveries in recent years is that environment triggers gene expression. Although most personal characteristics — everything from perseverance to memory — are influenced by our genes, they are not fully determined by them.

And then there's the issue of deliberate practice. In 1993, Swedish psychologist K. Anders Ericsson (now at Florida State University) proposed that expert performance was far more dependent on a long period of concentrated, deliberate practice than on innate ability or talent. Ericsson and his colleagues found that violinists and pianists whom faculty rated as the best musicians had devoted an average of more than ten thousand hours to deliberate practice by age twenty. "We attribute the dramatic differences in performance between experts and amateurs-novices to similarly large differences in the recorded amounts of deliberate practice," concluded Ericsson.

The paper, which suggested that practice time explained most (about 80 percent) of the difference between elite performers and committed amateurs, unleashed a torrent of research on the development of expert performance and has been cited in academic literature more than forty-five hundred times. In his 2008 best-selling book *Outliers,* Malcolm Gladwell touted what he called the 10,000-Hour Rule—the theory that ten thousand hours of appropriately guided practice is "the magic number for true expertise" in any field—which fueled one of the most persistent and influential pop-psychology claims of recent years. The concept's popularity—due to its meritocratic implication that almost anyone can excel at a chosen discipline if he or she tries long and hard enough—has been sustained by a slew of other best-selling books, including Daniel Pink's *Drive* (2009), Daniel Coyle's *The Talent Code* (2009), and Geoff Colvin's *Talent Is Overrated* (2010).

Neither Ericsson nor Gladwell went so far as to claim that innate talent was completely irrelevant to high-level success. "Achievement is talent plus preparation," wrote Gladwell, and Ericsson cautioned that effective training depended on the quality as well as quantity of practice time. But these finer points were lost amid Ericsson's continued insistence that "experts are always made, not born" and Gladwell's contention that diligent practice, more than talent, was the primary factor in the success of everyone from Bill Gates to the Beatles.

The influence of deliberate-practice theory has, at times, reached absurd levels. In 2010, commercial photographer Dan McLaughlin was so struck by the 10,000-hour proposition that he quit his job to attempt to become a professional golfer. The duffer, who started at 30 over par, told the BBC that "the goal is to . . . compete in a legitimate PGA tour event." Near the end of 2014, after 5,600 hours of practice, he'd managed to bring his handicap down to 3.1—respectable, although well above the 1.4 he needed to attempt to qualify for the U.S. Open tournament.

Subsequent research has shown that the answer is not nearly as simple as "practice makes perfect." For one thing, claims that deliberate practice nearly always trumps innate talent and intelligence conflict dramatically with the observations of teachers and others who work with gifted children. Many academics have derided Ericsson's "absurd environmentalism" and pointed out conceptual and methodological gaps in Ericsson's tests of his theory, which he later applied across several domains. More comprehensive research by others, including Wai and a team led by psychologist Brooke Macnamara, have found that the acquisition of expertise is, in fact, highly related to cognitive ability and that practice time explains only 20 to 25 percent of performance differences in chess, music, and sports. In another blow to deliberate-practice theory, Simonton, of UC Davis, has led several studies that found that people with the greatest lifetime productivity and highest levels of eminence actually required the *least* amount of time to achieve high-level performance. While practice is as important for prodigies as for other people, the time in which prodigies can amass the expertise needed for mastery in any given field is compressed. (A 2014 *New York Times* article on the research debunking Ericsson's deliberate-practice theory was titled "How Do You Get to Carnegie Hall? Talent.")

And yet, the evidence is stacking up that talent and practice are complementary, rather than oppositional, and far more intertwined than originally thought. All human characteristics, including the capacity and proclivity to deliberately practice, involve a mix of nature and nurture.

"Unfortunately, many people have an overly simplistic understanding of talent," says University of Pennsylvania psychologist Kaufman, who writes about intelligence and creativity in his *Beautiful Minds* blog for *Scientific American*. "In fact, there is no such thing as innate talent," Kaufman contends. "Gareth Bale wasn't born with the ability to score memorable goals. There are certainly genetic influences, but talents aren't prepackaged at birth; they take time to develop." In other words, high achievers are born, then made.

Some concept of talent may be necessary to help explain the development of high performance. But many researchers now argue for a more expansive definition of talent. Talent isn't just brainpower or acumen in a particular domain, they say, but any collection of personal attributes that quickens the development of expertise or improves performance given a certain degree of expertise. Simonton identifies four sets of characteristics—cognitive, dispositional, developmental, and sociocultural—and notes that a deficit in any one will lead to overall deficiencies. Tradeoffs can sometimes compensate (for example, someone with average intellect can become an overachiever if he or she is highly motivated), but these tradeoffs go only so far. "That's why exceptional talent is so rare," says Simonton.

David Lubinski, the Vanderbilt University psychologist who now codirects the Study of Mathematically Precocious Youth, adds *opportunity* to the constellation of personal attributes that lead to extraordinary performance.

"You could also include a million other things," says Kaufman, "such as physical features, social skills, and curiosity." There's also assertiveness, rebelliousness, self-confidence, and "grit," or the willingness to work hard. "The missing piece of the pie," Kaufman says, "undoubtedly includes other forms of engagement that don't feel as effortful as deliberate practice, such as play and flow."

If personal traits are as important as brains, so is another factor, and it isn't usually included on most researchers' lists—perhaps because, as a chance element, it can't be studied systematically. But this factor, Feldman says, is the most revelatory takeaway from his decades-long studies of prodigies.

When Feldman began working closely with exceptionally gifted children, his primary question was, What does it take to translate the raw material of innate intelligence into genius-level mastery and exceptional accomplishment? His conclusion, after thirty years of empirical research:

"Luck."

According to Feldman, the path from supersmart kid to world-changing adult depends mostly on what he calls "the co-incidence process." Feldman's research (now backed up by others) makes it clear that the circumstances have to be just right for talent to flourish. "From the starting point of innate, natural ability," Feldman says, "specific talents tend to require specific environments very well suited to their development."

For instance, the SMPY subjects were a fortunate bunch of kids from the start. Without encouragement from parents or teachers, they likely wouldn't have taken the SAT early. Their advantages continued to compound after they were identified as "exceptionally gifted." Unlike many—perhaps most—gifted children, they gained access to unusually rich learning experiences. These included special attention from schools and teachers and invitations to hyper-intensive summer programs, such as the ones that Zuckerberg, Gates, Jobs, and Germanotta attended, where they could gorge themselves on a year's worth of math or science or literature in a few weeks.

Two recent papers published in the *Journal of Educational Psychology* found that among young people with high ability, those who were allowed to skip a grade, enroll in special classes, or take college courses in high school were significantly more likely to earn PhDs, publish academic papers, develop patents, and pursue high-level careers than their equally smart peers who didn't have these opportunities.

Everyone's heard the bright-kid-overcomes-all anecdotes. But the bigger picture, based on decades of data, shows that these children are the rare exceptions. For every such story, there are countless nonstories of other gifted children who were unnoticed, submerged, and forgotten in homes and schools ill-equipped to nurture extraordinary potential. In those environments, David Hahn–type outcomes are far more prevalent.

"There has to be an almost uncanny convergence of certain things lining up," says Feldman. "The gifted child must be exposed to a field or art, someone must observe their interest and act upon it, and the timing and cultural context and available technology must all be right. Parents and teachers must have the resources and work hard to connect the prodigy with the right mentors or coaches, if the child is going to achieve any significant portion of innate potential."

"This is not a trivial point," says Winner. "Because it indicates that, of the [possibly] millions of children who are born with the potential to propel themselves to mastery, only a tiny portion are ever given a chance, due to accidents of fate. Imagine if Taylor had been born as an Aborigine in the Outback in Western Australia. There would be no technology, no environment, no mentors, no cultural context that would have matched his interests and abilities."

"Even if Taylor would have grown up on the Upper East Side of Manhattan," says Feldman, "things likely would have not worked out so well for him." He would have been born into a culture that supported high achievers, but it's hard to imagine where he'd have dug his holes or shot off his rockets—or found neighbors who would tolerate his explosions and his glowing radioactive goodies. His parents might have had more financial ability to support his endeavors, but they likely would also have had more competing distractions—as would Taylor. Money can buy time, but when it comes to parenting, it often does not. Money can also buy a top-notch education, but prestigious prep schools are not set up to indulge exotic talents unrelated to bagging a spot at an Ivy League university. Formal schooling is just one piece of the prodigy puzzle, which also includes parenting, personal characteristics, social/emotional development, family aspects (such as birth order, gender, and traditions), access to resources, and historical forces and trends. When all those things happen to be in coordination and are sustained for a sufficient period, a child born with extraordinary potential can bloom. When one or more elements are missing, inborn talent is more likely to wither.

"That's a lot to get right," says Feldman. "And because of that, only a tiny portion of would-be prodigies go on to become eminent, creative adults."

16

THE LUCKY DONKEY THEORY

IN A WAY, the amateur nuclear fusion movement began at the intersection of science and science fiction. In the mid-1990s, an electrical engineer and aspiring science-fiction writer named Tom Ligon heard that the physicist Robert Bussard was living and working just two miles from his Virginia home.

"Some people still think Bussard is a fictional character," Ligon says, "but it turns out that he was quite real." In the mid-1950s, Bussard worked at Los Alamos in the Nuclear Propulsion Division, designing nuclear rocket engines. He coauthored two books on nuclear-powered flight and in 1960 proposed the "interstellar ramjet," which would scoop up interstellar ionized hydrogen and funnel it into a nuclear fusion reactor whose output would propel a spaceship forward. Though Bussard thought his design was at least two hundred years from feasibility, the interstellar ramjet (also called the Bussard ramjet) became a fixture in science fiction, used to propel space travelers in Larry Niven's and Poul Anderson's novels and the Starfleet starships in the Star Trek universe.

Bussard went on to become assistant director at the Controlled Thermonuclear Reactions Division of what was then the U.S. Atomic Energy Commission. The division had settled on magnetic confinement as its mainline fusion program, but Bussard saw more promise in a fusion reactor he called a polywell (a combination of *polyhedron* and *potential well*), which overcame some of the inefficiencies of the fusor. Bussard's reactor used a negatively charged plasma field instead of a negatively charged wire grid to attract and accelerate the positively charged ions. Bussard started a small tech company, funded by the military, to pursue his polywell design.

Ligon dropped by with his résumé. "Nobody was there," Ligon remembers, "but the sign . . . declared that it was the Energy/Matter Conversion Corporation. I slipped my propaganda under the door, smiling as I got the connection to Einstein's famous formula."

Bussard called him back, and Ligon went to work for EMC2, which built several polywell prototypes between 1994 and 2006. With the final prototype, Bussard felt that he'd solved the remaining major physics problems and he reported exponential efficiency improvements. But as the Iraq War consumed the military's resources, the military defunded polywell research. Bussard raised private capital to build a polywell power plant.

Bussard passed away in 2007. "He was convinced when he died," says Ligon, "that he had achieved a significant breakthrough." U.S. Navy researchers may have been similarly convinced, since they restarted the polywell program after Bussard's death and brought it to Los Alamos. Taylor and Willis are among the few outsiders who've seen the next-generation polywell machine; since their visit to the laboratory, the project has been classified.

In 1998, Ligon built an almost-functional Farnsworth-Hirsch reactor and brought it to a meeting of amateur high-energy scientists at Richard Hull's home near Richmond. "While I was standing in awe of Richard's mighty Tesla coil," Ligon says, "the others were falling madly in love with the idea of building their own tabletop hot fusion reactors."

The next year, Hull's fusor became the first outside a research laboratory to achieve a verified nuclear fusion reaction. Others in the community followed, and founded an online forum to share resources with people like Willis, who in 2003 became the tenth person to build a working reactor. In addition to Fusor.net's role as an information clearinghouse, the forum has become the de facto verification body for claims of amateur nuclear fusion success

Fusor.net's fusioneers list has three levels. Scroungers are just starting out, gathering components and/or assembling parts. Plasma Club membership requires evidence of plasma production. Neutron Club applicants must provide rock-solid proof of fusion in the form of images and a full data disclosure regarding setup, conditions, and neutron-detection systems. "Many of the people who get the idea to build a fusion reactor have that long-shot sci-fi edge," Willis says, but the technical challenges quickly weed out those who aren't serious about mastering nuclear theory and engineering. "You can't fake your way into the Neutron Club," says Willis.

"Where people usually fall short is in their neutron-detection methods," says Hull. There's a lot of back-and-forth questioning and answering, which serves as an informal but tough peer-review process. In early 2014, for instance, a thirteen-year-old named Jamie Edwards, a student at Penwortham Priory Academy in England, announced fusion success. British tabloids and even David Letterman jumped the gun and ran with the story—but the consensus in the Fusor.net community was that Jamie had not decisively achieved fusion. Forum members suggested detailed steps he could take to bring his verification methods up to a convincing level. The youngster accepted the suggestions gratefully and continued to work on his fusor.

Taylor added his name to the Scroungers list soon after Willis, who helps administer the Fusor.net site, suggested that Taylor build a fusor. Taylor became an increasingly frequent presence on the site's technical forums. "Everyone was incredibly generous and willing to give advice and let me bounce ideas off them," Taylor says. "There's not too many sources of know-how for some things, like vacuum issues, for instance. Vacuum is almost a black art; I mean, who goes to school for it?"

But at first, none of the forum contributors had any clue that they were communicating with a twelve-year-old.

"No, it never occurred to me until Carl told me he'd talked with Taylor on the phone," says Hull. "I was amazed. By the level of his communications, I thought he was someone much older, with a great deal of academic physics education under his belt."

Hull hosts the High Energy Amateur Science (HEAS) Conference each autumn. The invite-only event is a weekend-long swap meet and series of demonstrations and experiments involving radiation, lasers, and extremely high voltages. Taylor wasn't able to make it to the HEAS gathering the year I attended, but Willis hasn't missed one since 1993. He says he felt immediately at home at this semi-underground gathering. "Just knowing there were people who were interested in the same things I was, that was huge," Willis says. "I also have to admit that the element of danger appealed to me."

"These are not wine-and-cheese parties," Hull had told me over the phone. Indeed, when I pulled into the lot next to Hull's house midway through Saturday morning, my first impression was that I'd arrived at a sort of cerebral, hands-on car club, a flea market for high-end nerds. The yard was filled with vehicles, their trunks and tailgates open, and tarps and tables stacked with lasers, electron multipliers, vacuum tubes, and plenty of radioactive rocks. I overheard a conversation that began with the question, "Any idea where can I get some terbium-activated gadolinium oxysulfide?"

Hull wasn't hard to find. In his late sixties, with the potbelly of a man who's too busy tinkering to put much effort into taking care of himself, he's an electronics engineer by trade, though his career rarely intersects with his hobbies. Neither does his wife, who stayed inside the house almost the entire weekend. Hull told me, in an accent that matches Slim Pickens's character's in *Dr. Strangelove,* his rationale for hosting the gathering: "I try and take what would normally be a lot of lone-wolf, clever folks who are spread across the continent and force them to mix and share their thoughts, ideas, and materials once a year."

Near Hull's workshop, someone was wiring a very large hand-built Marx generator, which produces extremely powerful pulses of both voltage and current by using multiple capacitors that are charged in parallel and then discharged in series across spark gaps. "That," said Frank Sanns, a self-employed science consultant who was watching the setup, "is by far the most dangerous thing here." Marx generators produce electromagnetic pulses (EMPs) that can fry electronic devices—and anything else that gets too close. "Better take a step back and hide your digital camera and gonads when it comes on," Sanns said. "Last year it took out a couple of cameras and phones."

Tinkering was an essential part of life for much of the world's population until very recently, and budding scientists and engineers learned by doing and fixing. During his teenage years, Richard Feynman taught himself how to repair radios, developing a business that helped support his family through the Great Depression and nourishing a curiosity that led to his Nobel Prize–winning brainstorms.

The post–World War II years were the era of Big Science, but they were also a boon to small-scale science—much of which was inspired by the space race and the Cold War. Hull got into experimental chemistry as a teenager in the 1960s. "Back then," he told me at the gathering, "you were free to blow off an arm or leg making steel rockets. We used to go on a bus across town and pick up fifty or sixty pounds of ammonium nitrate. Try doing that now!"

As part of its Atoms for Peace program the Atomic Energy Agency distributed free radioactive materials to amateurs, hoping to inspire the generation that would push America further into atomic age. "I remember," Hull said, "when the mailman actually brought cobalt-60 to my front door!"

It was a profitable time for businesses whose products met the demands of do-it-yourself scientists. The A. C. Gilbert Company thrived with its Erector sets (called Meccano outside the U.S.), telescopes, and atomic energy labs that included a Geiger counter and other instruments and even small radioactive sources. Heathkits, electronic kits marketed by the Heath Company, helped would-be electrical engineers learn the workings of radios, televisions, and other electronic gear by self-assembling them. "The kits taught Steve Jobs that products were manifestations of human ingenuity, not magical objects dropped from the sky," writes Leander Kahney in his book *Inside Steve's Brain*. Jobs credited his Heathkit explorations with giving him "a tremendous level of self-confidence, that through exploration and learning one could understand seemingly very complex things in one's environment."

The lure of explosive chemistry was probably the most potent recruiting tool midcentury science had to offer. Hewlett-Packard cofounder David Packard, Internet pioneer Vint Cerf, and author/neurologist Oliver Sacks all credit their interest in science to childhoods spent blowing things up with chemistry sets. Intel cofounder Gordon Moore was eleven when he exploded his first chunks of homemade dynamite. "There's no question that stinks and bangs and crystals and colors are what drew kids—particularly boys—to science," says Roald Hoffmann, who won the Nobel Prize in Chemistry in 1981.

The trend away from do-it-yourself science began in the 1980s, says Bob Parks, author of *Makers: All Kinds of People Making Amazing Things in Garages, Basements, and Backyards*. As cheap, well-sealed electronic gadgets became easier and cheaper to replace than to repair, interest in building things and taking them apart plummeted. By the turn of the millennium, says Parks, that "'Hey, let's just make stuff' mentality was mostly washed out of mainstream America."

In 2001, *Scientific American* discontinued its Amateur Scientist columns; editors said readers were no longer interested in hands-on science. The magazine then published an essay, titled "R.I.P. for D.I.Y.," by George Musser, who lamented the loss of much of "the mentoring and serendipity" that communities of amateur scientists offered.

Today you'd be hard-pressed to find a child who is motivated to get under the screen of a smartphone to figure out what makes it light up

— and you'd be even harder pressed to find a parent who would encourage it. That's made it less likely that today's kids will have the kind of formative experiences that Feynman, Moore, and Jobs had.

Parental concerns about safety (which are often legitimate) can stifle young scientists, as can restrictions on chemicals, spurred by concerns about terrorism and the epidemic of crystal methamphetamine made by amateur cooks in clandestine laboratories.

In schools, science labs that benefited from spending sprees in the wake of Sputnik have suffered decades of neglect. The chemophobia that stifled home science has penetrated classrooms, where a fear of lawsuits makes teachers wary of letting students perform their own experiments. Kids who attempt science experiments on school grounds can find themselves handcuffed and charged with a felony, as a Florida high-school student found out in 2013 after she induced a chemical reaction that popped the top off a water bottle and produced some smoke. Though no one was hurt and nothing was damaged, the school expelled sixteen-year-old Kiera Wilmot, who had good grades and a perfect behavior record. *Rocket Boys* author and NASA engineer Homer Hickam came to Kiera's aid after hearing her story, and he presented her with a scholarship to the United States Advanced Space Academy.

Science without interactive labs or projects that relate to students' real-world experiences just isn't that much fun and may be contributing to the declining interest in STEM careers. Thirty years ago, the U.S. ranked third in the number of science and engineering degrees awarded in the eighteen- to twenty-four age group; today the U.S. ranks seventeenth, according to the National Science Board. "Human beings learn best by exploring or investigating, not by ingesting and swallowing facts and figures," says author and education-reform activist Nikhil Goyal, who advocates bringing back shop classes and introducing project-based learning, in which students probe real-world problems collaboratively. "Learning," Goyal says, "should be messy."

Those who are motivated to do their own science say that, even as the Internet made it easier to *learn* how to do things, the hyperfocus on safety and security often made it harder to *actually* do them. "The list of things that are too dangerous, too corrosive, or too explosive seems to grow every day," says Hull. The Porter Chemical Company, maker of the popular Chemcraft labs in a box (each of which had enough liquids, powders, and beakers to conduct more than eight hundred experiments), closed its doors in the 1980s amid liability concerns. Most states restrict materials that could be used to make illegal fireworks or meth. Texas has gone so far as to require buyers and sellers of Erlenmeyer flasks to obtain permits and register the equipment.

Bill Nye, who hosted the award-winning PBS series *Bill Nye, the Science Guy,* believes the restrictions have gone beyond reason, to the level of paranoia, and that the eventual negative consequences will far outweigh the benefits. "People who want to make meth will find ways to do it with or without an Erlenmeyer flask," Nye told *Wired* magazine's Steve Silberman. "But raising a generation of people who are technically incompetent is a recipe for disaster."

"Even in the 1980s when I was a kid," says Carl Willis, "I could stop into the pharmacy and the pharmacist would order saltpeter for me, which I could use in my explosive experiments. The lack of freedom to do those kinds of things has made it really hard for young people to find something that captivates their interest, which only happens when they are allowed to interact with the material world, the physical universe, in a way that's new to them."

Then again, some young experimenters who have no interest in physically taking apart their video-game consoles or smartphones have taken their tinkering to the virtual world. They're hacking and modding games, manipulating data, bending programs to their will.

"This isn't something that should be dismissed," says Shawn Carlson, a MacArthur Foundation genius-grant recipient who wrote the final Amateur Scientist columns and who now trains teachers to inspire children to love science. "Most kids just want to play the games. But some want to build something purely digital that's ambitiously new and different."

Pioneering computer scientist Ted Selker disagrees. "Something important is lost when kids don't have—or don't take—the opportunity to explore the world with their hands." Known for inventing the TrackPoint device for IBM's ThinkPad and for his user-interface innovations at MIT's Media Lab, Selker believes there's no way for someone to fully develop creatively by just staring at screens and tapping at keyboards. "Every time we touch a piece of bendy aluminum or soft copper, our brain builds a library of the physicality of that object, and the possibilities for it," Selker says. "The ability to learn conceptually and not just procedurally is created by the process of taking things apart and building things; that's how we develop the intuition to make useful and creative connections."

This sentiment isn't just poetic nostalgia, says physicist Steven Cowley, who leads Britain's nuclear fusion program. Cowley believes that the lack of hands-on experiences is actually holding back innovation. "We have a big uptick in engineers and physics students coming in, but we don't have good experimentalists with a feel for what's going on, because they've lost the practical skills." Cowley says he now keeps an eye out for engineers who are sons or daughters of farmers. "They've grown up getting the spanner out and learning to fix things, and that gives them a problem-solving perspective that you can't get any other way," he says.

"That's what's most impressive about Taylor," says Cowley. "Not that he understands this stuff that practically no one else understands, but that he can build it. It's one thing to understand physics at the age of thirteen, quite another to then apply the theory to a machine that you hand-build, largely by yourself, of begged-and-borrowed and adapted parts."

Despite the roadblocks, there are signs that a small but countervailing trend is building and that interest in hands-on amateur science is on the verge of a comeback. The phenomenon is emerging via several different paths. Citizen scientists have become the workhorses of crowd-sourced data collection and have made major discoveries and contributions in the fields of astronomy, environmental science, energy, and aviation. Top scientific journals have opened their peer-reviewed pages to research papers coauthored by self-taught experts such as Forrest Mims III, who discovered that LEDs (light-emitting diodes) have the ability to both emit and sense light.

At the same time, a new generation is intent on seizing the traditions of do-it-yourself science and remaking them in its own tech-focused image. The grassroots maker movement encourages hands-on and participatory creation, especially in pursuits such as electronics and robotics.

The movement had its roots in the sixties and seventies with publications like Stewart Brand's *Whole Earth Catalog* and organizations like Silicon Valley's Homebrew Computer Club, whose hobbyists largely launched the personal-computer revolution. But maker culture really took off in the first decade of the twenty-first century, when hacker spaces, fab labs, and other maker spaces began popping up in university towns, allowing geeky creators to share skills, ideas, and tools. Maker

Faires, robotics fests like Dorkbot, and media such as *Make* magazine and *Boing Boing* have boosted interest and a sense of community and helped translate the hacker ethic to the nonvirtual world. Suddenly, it's cool again to get your hands dirty.

The high-energy amateur science community is an unusual maker subculture. Whereas most amateur scientists are into either theory (they want to know why) or building (they want to know how), HEAS people need both a command of complex atomic physics theory and the hands-on skills to do precision engineering work.

Like many at Hull's HEAS gathering, Sanns, the science consultant, has ideas for pushing nuclear fusion forward. "I've recently had a eureka moment," he told me. "All the approaches so far have been to hit it harder, ramp things up, smash them together, hope for the best. I've been thinking about the strong nuclear force and electrostatic repulsion and think I might have a better way."

The long-prevailing approach to magnetic confinement fusion is that bigger can only be better. The need for such massive scale — ITER is by far the biggest of all the Big Science projects — increases the complexity and the range of problems both scientific and political. Big Science has in many cases yielded big results, "but big science has the special problem that it can't easily be scaled down," physicist Steven Weinberg writes. If ITER succeeds, no one will regret betting big on it; if it fails, the cacophony of "I told you sos" will likely drown out any suggestions of taking on something this bold for a very long time. In the uncertain meantime, the question has to be asked: Is Big Science still the best way to get things done, or is it buckling under its own complexity?

Back in 1911, Rutherford's gold-foil experiment, which resulted in the discovery of the nucleus, was financed by a seventy-pound grant and consisted of arranging a few items on a tabletop. Physicist Robert Millikan made his groundbreaking discovery of electrons by using a perfume atomizer to squirt a mist of oil and clicking his stopwatch to time how quickly the droplets fell through an electrical field.

"As the effort to understand the world has advanced, the low-hanging fruits (like Newton's apple) have been plucked," wrote George Johnson in the *New York Times*. Since Rutherford's and Millikan's discoveries, big breakthroughs have come at an ever-higher cost and have required ever more complex machines and supporting administrations. The Large Hadron Collider, seventeen miles in circumference, produced electronic data equivalent to billions of Millikan's notebooks to verify the existence of the Higgs boson and the Higgs field, which gives fundamental particles their mass.

The massive investments in ITER don't make it too big to fail; they make it too big to abandon. And yet, tokamaks like ITER may not represent the best long-term approach to fusion. "Tokamak technology is as twenty-first-century as you can get," says plasma physicist Ron Phaneuf, "but you are still heating water to make power."

Meanwhile, a few startup companies are exploring radically different approaches, such as aneutronic fusion, which delivers electricity from the fusion process without using a turbine, and magnetized target fusion, a hybrid of magnetic and inertial confinement fusion. If fusion energy is still waiting for its big idea—a fresh approach that can leapfrog the decades-long incremental push toward more complexity—is it likely to emerge from a massive scientific bureaucracy? Or could it spring from a highly motivated individual, a small company, or one of the collaborations that the new model of participative amateur science is constantly spawning?

The fact that fusion reactors are now being built relatively cheaply by crowdsourcing and online collaboration is consistent with the way innovation is evolving in the twenty-first century. While the amateurs' experiments aren't nearly as advanced as those done in multibillion-dollar facilities, it's conceivable that developers of these homebrew reactors could play a vital role in moving fusion forward, as citizen scientists have in other realms.

Hull calls it the "lucky donkey theory," the idea that many scientific advances are achieved by creative underdogs who lack the resources and high technical standards of established laboratory-based researchers but make up for it in ingenuity and a higher tolerance for risk.

Carl Willis, with his nuclear engineering job and his high-energy hobbies, has a foot in each world. "The story of the underdog topping well-heeled opposition can be a seductive cliché," he says. "But I don't think it's as far-fetched with physics as the industry leaders would have us believe. There's a certain stodginess that impedes innovation, and from what I've seen, a lot of people in the big labs don't have much of the creativity and enthusiasm that lead to big breakthroughs."

As Hull beckoned the HEAS crowd into his workshop for a demonstration of his fusor, fusioneer Edward Miller told me that he recently flew to the University of Texas to conduct fusion-related experiments.

"We got some money together and used their lab's lasers and accelerators," Miller said as we approached the door. "We were trying a lot of things, going for big numbers, using pressure instead of temperature. We blew deuterium into buckyballs and hit them with x-rays, electrons, protons, lasers. We didn't get any neutrons, but we got tons of data. I don't think we made any breakthroughs, but I think we pretty conclusively ruled out a few things.

"Hey" — he shrugged and pulled open the workshop door — "some people work on cars."

17

TWICE AS NICE, HALF AS GOOD

TAYLOR WAS NOT ONLY twice blessed—with his intellect and his parents—but thrice blessed, with a school environment that could accommodate his personality and learning style. And yet, St. James Day School, which had worked so well for Taylor, wasn't working nearly as well for Joey by the time he reached the upper-elementary level.

Joey was as smart as Taylor—in fact, he scored consistently higher on every standardized test—but he didn't have his brother's luck of the draw when it came to teachers, or his talent for self-advocacy. He had shown an early aptitude for math, and he wanted to tackle more advanced material. "But unfortunately," says Dee Miller, head of the school, "at the time, we had a teacher who was not as flexible as what we would want."

Kenneth and Tiffany went in and asked the teacher if he could give their son some more challenging work. But, according to Kenneth and to Miller, the teacher told them, "We will only give him more of what everyone else is doing."

"Teaching to the middle doesn't work, never has," says Darren Ripley, who taught upper-level math to both brothers. "To say we're all at an Algebra Two level just because we're in the tenth grade just isn't accurate. It also isn't effective."

Joey's learning style was similar to Taylor's learning-through-talking style, though quieter. "I learn through testing," he says. "I go to the lectures, take the tests, but I never study. With some stuff, it takes being in the test to draw the connections. I usually come out knowing things I didn't know going in. I think I'd psych myself out if I tried to study; I'd be more anxious."

If my son told me that, I'd be skeptical. But with Joey, the proof was on the paper (and the shower door). When he took the ACT college-admissions test in the seventh grade (he didn't want to take it and stayed up the night before until 1:00 a.m., in protest), he scored in the top 99.9th percentile in math; in other words, he was a math prodigy. Even in fourth grade, he was clearly mathematically gifted, and yet his teacher insisted that he do the same work his classmates were doing — even though he had already mastered it.

"He was miserable in class," says Miller, "especially his last year here."

Nell's weekly yoga sessions, which were a big hit with Taylor's and Joey's classmates, came to an end when Taylor was in the sixth grade. After Nell had a round of chemotherapy, the family traveled with her to an alternative-medicine detoxification clinic in Mexico, where she bounced back briefly. But once she got home, her decline accelerated. In desperation, Tiffany brought her to another clinic in Arizona, but shortly after they returned, Nell passed away. At the memorial service, Taylor sang "Amazing Grace."

"Everyone missed her coming to school to do yoga with us," Taylor's friend Ellen Orr says. "And for a while, Taylor and Joey got really quiet. Then Taylor started getting more excited about this idea he had, to use nuclear fusion to make medical isotopes."

Nell's death affected her older grandson in profound ways, some of which would become apparent only years later. Like any child, Taylor reacted with grief and anguish. But he also had another, more unusual response — a sense of defiance that led to a spark of inspiration and a vision of a future in which he would change the world through nuclear fusion.

But the success of his nuclear reactor would depend, in large part, on achieving the extremely high voltages and vacuum pressures that had eluded him in his attempt to build his particle accelerator. Unfortunately, the specialized high-voltage feedthroughs and high-precision vacuum flanges and seals that physics laboratories use were still out of his reach financially.

"I remember," Willis says, "how much back-and-forth there was on the Fusor.net forum about how Taylor could get good vacuum seals on the cheap. He had good ideas, and it seemed like he was making pretty good progress, considering his constraints."

When his grandmother's house went up for sale, Taylor moved his lab to the Little House, the two-story shed in his backyard. He searched the Internet for low-cost parts that he could adapt for his fusor's reaction chamber, its power supply and vacuum pump, and its various fittings and electrical components. To finance his purchases he started a business assembling and selling small radioactive check sources, which are used to confirm the proper operation of instruments such as Geiger counters. Taylor made the sources by crushing up small, nonhazardous amounts of uranium ore and thorium lantern mantles, then inserting them inside small test tubes and sealing them in epoxy glue. The business experienced a mini-boom when an environmental contractor bought forty sources for training purposes.

Working in the Little House and at the Coke plant, Taylor started to assemble his machine out of the hodgepodge of secondhand equipment he'd collected. But the low-budget approach turned out to be frustrating; parts often didn't fit together, and without access to a proper machine shop, he often ended up making additional purchases and retrofits.

Taylor had thought that once he got to middle school, he'd find well-outfitted science laboratories and "people who knew stuff," as he put it, who could help him with his projects. "But the labs were outdated," says classmate Ellen Orr, who continued, with Taylor, to Texas Middle School. "And no one knew anything about the stuff he was talking about or took an interest in helping him move ahead."

Tiffany and Kenneth had worried about whether Taylor would be accepted by the other students at Texas Middle School. But he quickly fell in with a crowd who appreciated his somewhat squirrelly and decidedly nonmasculine nature. Socially, seventh grade was working out for him. But intellectually, it was a washout.

"All they could offer was regurgitation," Taylor says. "There was no exploration or discovery; it was just 'fill in the blanks.' And it moved sooo slow."

Forcing a highly gifted child to work at the average student's pace is, Lisa van Gemert told journalist Susan Freinkel, "like forcing an adult to play an endless game of Candy Land." Gemert, a gifted-youth specialist for American Mensa, says these children will ultimately drop out, either literally or figuratively. "They may turn their frustration inward and become depressed and self-destructive, or turn it outward and mouth off to teachers and stir up mischief."

Taylor quickly found himself at intellectual odds with his science teacher. "She was incompetent," says Ellen Orr bluntly. In his Integrated Physics and Chemistry class, Taylor butted heads incessantly with the instructor. "She was teaching from the standard physics textbook, which was outdated," says Ellen. "She'd ask a question and Taylor would cite newer evidence that conflicted with the textbook answer." For instance, teacher and textbook taught that atoms had only three parts — protons, neutrons, and electrons. But on quizzes and tests, Taylor said, "I'm not putting that answer down, 'cause it's not right."

"He'd launch into a lecture on the six flavors of quarks [the fundamental particles that combine to form protons and neutrons]," Ellen says. "I still remember the weird names: up, down, strange, charm, bottom, and top. And while he was at it he'd want to explain the forces that kept them together. He'd be like, 'New research says . . .' and everyone would start getting fed up."

The fact that he was right didn't make it any easier for Taylor, who had been used to getting his way when it came to education (and just about everything else). But his school's educators didn't have the skills to recognize how he learned best or the resources to accommodate it if they did. With so many students struggling to grasp the basics, there was no time to let a kid prattle on about technetium-99 or the esoteric behaviors of baryons and mesons.

Midterm exams came and went, and Taylor stuck to his guns, refusing to put down technically wrong answers. "Newton's understanding of gravity was limited!" he lectured his teacher one day. "Einstein figured out that gravity is not a force, but a curvature."

"I was sitting in the room after class," says Ellen, "when the teacher called Tiffany and told her Taylor was failing."

Taylor and Ellen were reaching another of Kerr's stages of vulnerability. Adolescence is a time of great change, when boys and girls need to develop effective coping mechanisms before the additional stresses of high school. All too often, they don't. Puberty is, of course, a minefield for both sexes, and middle schools have not solved the riddles of how to guide students through it. The problem is not unique to North America. Dr. Stephen Tommis, former director of the Hong Kong Academy for Gifted Education, says, "Without genuine understanding and the right kind of emotional and educational support, [gifted children] sometimes start to fall apart almost as soon as they reach upper primary or junior secondary school."

"The peer pressure to be obedient to gender roles intensifies, especially for gifted kids," says Kerr. "Girls learn that it's okay to be smart if you're pretty and popular. Suddenly they're no longer reinforced by their talents but their ability to achieve a boyfriend." Gifted boys are particularly vulnerable if they become aware that their peers associate academic achievement with compliance. "Boys see noncompliance as a sign of masculinity," says Kerr. "They don't want to be the teacher's pet. So gifted boys will often underachieve as a way of establishing masculinity with other boys."

Numerous studies confirm a sad finding: intellectually gifted students typically have little good to say about their schooling. "They're usually bored," says Winner, "and they tend to be highly critical of their teachers, who, they feel, know less than they do." When ability grouping or other acceleration isn't an option, says Winner, the best-case scenario is a teacher who recognizes a student as gifted and lets the child learn independently. The worst-case scenario is a teacher who fails to recognize giftedness and brands the child as unmotivated or hostile.

Taylor had fallen out of the nurturing cocoon of his elementary school, where the system could flex to fit a student's needs, and into the harsh "industrialized model" of education, which demands that students fit into the system. Education and creativity expert Sir Ken Robinson, whose TED talk "How Schools Kill Creativity" has been viewed more than thirty million times, notes that our educational system developed from a factory culture. "We still educate children in batches, with ringing bells, days divided into forty-minute bits, children separated into age groups." The system punishes those who slow the assembly line and creates a culture of conformity. "We are educating people out of their creative capacities," says Robinson. "School is not challenging them, firing up their imaginations."

But if gifted kids are so smart, won't they find their own way? Those who work with gifted and talented children say that's among the biggest misconceptions. "There's this idea that they don't need much help, and it's clear that's not the case," says Vanderbilt researcher David Lubinski. In one long-term study, Lubinski found that supersmart kids will often do "just well enough" if they're stuck in regular classrooms that can't accommodate the rapid rates at which they learn; they won't come close to their full potential.

The exceptionally talented often become invisible in the classroom. "When students enter on day one having already mastered the material, teachers immediately shift to the underachievers who are struggling," says Lubinski. Fast learners tend to escape criticism, but if they lack challenging coursework that incorporates their interests, they lose motivation. The dropout rate among students in the gifted range may be as high as 5 percent, but more commonly, potential high achievers who are unengaged just "get through" the school day. One study, by the University of Iowa's Belin-Blank Center, found that nearly half of gifted kids are underachieving.

In America, an academically or artistically gifted child's ability to develop his or her talents depends, to a tremendous extent, on where that child lives. A lucky kid who's stagnating in a local school may be able to transfer to a more stimulating environment if she lives near a magnet school or a town that offers advanced content. "An unlucky gifted kid," says Winner, "usually just starts to hate school."

Tiffany and Kenneth could see that neither of their children was getting what he needed from his school. But with this issue, the couple's problem-solving abilities had hit a wall. On both sides of the state border in Texarkana, the public schools do not excel. And among local private schools (most of which lean heavily on the Bible and lightly on intellectual development), St. James was an outlier, with no counterpart in the higher grades.

"This is a town," Dee Miller says, "that doesn't really value education."

Nikita Khrushchev probably did more to boost gifted education in America than anyone else. "The Sputnik launch provoked all kinds of support and programs for talented kids, and acceptance of approaches like acceleration and grade-skipping," says Rena Subotnik, who directs the Center for Gifted Education Policy at the American Psychological Association.

The boom years for bright kids continued into the 1970s. Options expanded with the growth of talent searches, school-pullout programs, scholarships, summer residential programs, math and science Olympiads, and early college admissions. Suddenly, even basket-case school districts were throwing together gifted-and-talented (G&T) programs. Some were well conceived and executed; others were slapdash or wildly experimental. But all were hobbled by the fact that they could draw on very little quality research as to what actually *worked* in gifted education. That research eventually started flowing as funding arrived and specialized academic centers began to produce authoritative studies that shed new light on this segment of the population.

But providing gifted children with an education that matched their abilities hit an ideological roadblock in the late 1980s and early 1990s.

America's Sputnik inferiority complex dissolved with the breakup of the Soviet Union, and academics began searching for a wider, more democratic conception of human excellence. Specialized education for the academically talented came under attack, swept up in a tide of antielitism.

"The word *gifted* itself didn't help," says researcher Paula Olszewski-Kubilius, "with its implication that the kid got something other people don't have, and he should just be satisfied with that."

During the 1990s, programs for high achievers were cut back or eliminated. Even as a growing body of research confirmed the effectiveness of practices like ability grouping, acceleration by subject, and grade-skipping, these low-cost or no-cost solutions fell out of favor due to concerns about inequality and fears — which were not supported by research — that acceleration hurt children socially and hurried them out of childhood.

"The idea of singling out a few kids for special treatment offended egalitarian sensibilities," says Winner. The discomfort level rose if those kids were well-off whites who already had access to the best schools and widest opportunities. Outspoken researchers such as Samuel Lucas, a Berkeley sociologist who studies inequality in education, raised politically fraught questions about whether schools should be in the business of segregating elite students and thereby exacerbating inequality — especially inequality linked to race and class. Lucas wasn't necessarily against efforts to identify and serve gifted children, but he warned — with significant justification — that such programs would inevitably be gamed by well-off parents who would find ways to get their children in and then fight to ensure that the gifted set got better teachers and curricula, newer technology, and more funding. It's difficult, says Lucas, not to "end up with another system for those at the top to reinforce that they belong there."

Advocates for gifted children say concerns about elitism miss the point: kids of *all* abilities deserve an education that identifies each student's needs and gives that student the support to develop his or her gifts, whatever they may be. Nevertheless, dedicated G&T programs started declining and disappearing, first in liberal college towns such as Cambridge, Massachusetts, and Ann Arbor, Michigan; they were more likely to persevere or enjoy early comebacks in more conservative suburban districts. "The more PC the community, the less likely it is to have a dedicated program," says Winner.

In university towns, dedicated gifted programs were already stressed by über-educated parents' demands for access. In the mid-1990s, when the Ann Arbor public schools were exploring alternatives to serve gifted students, the district sent out a survey to parents, remembers Lee Ann Dickinson-Kelley, assistant superintendent of instruction and student support services. Among other questions, parents were asked about their children's talents.

"What percentage of the respondents do you think thought their child was gifted?" Dickinson-Kelley asked me. "Ninety-nine percent," she answered. "I'm not kidding you."

After the turn of the millennium, a long-overdue focus on the nation's underachievers further overshadowed the educational needs of high achievers (and many other aspects of education). "Odd though it seems for a law written and enacted during a Republican Administration," wrote *Time* magazine's John Cloud, "the social impulse behind No Child Left Behind is radically egalitarian. It has forced schools to deeply subsidize the education of the least gifted, and gifted programs have suffered."

When the law took effect, in 2003, the federal government and the states began pulling funding away from G&T programs. Illinois cut sixteen million dollars, New York fourteen million, Michigan four and a half million. Although some states maintained fairly large G&T budgets (Iowa's was close to thirty-six million in 2014), about a dozen states now spend nothing at all on gifted programs.

Although the Obama administration tweaked some aspects of NCLB, including its name, the focus is still on achieving grade-level proficiency. "They call it Race to the Top," says NAGC director of public education Jane Clarenbach, "but it isn't really about the top at all; it's about achieving grade-level proficiency, which does nothing to help gifted kids. We believe it's the responsibility of public schools to provide appropriate education to all children."

The post-Sputnik decades of increased attention on gifted kids led to a staggering growth in highly educated people who produced science and engineering innovations that have extended both the quality and length of people's lives, created tens of millions of jobs, and fueled much of the West's economic growth.

Now, in an era in which prosperity depends ever more on the

intellectual capability of a nation's population, our educational system has failed to nurture and develop the talents of its most promising students. The rate of increase in the number of U.S. doctorates has fallen dramatically since its high in 1970, and the most recent round of international test results showed a continued decline in the performance of top U.S. students compared to their global counterparts.

Fifteen-year-old Americans ranked twenty-fourth and twenty-eighth, respectively, in reading and science. While most other nations' scores had risen since the previous test, in 2009, U.S. performance had changed little. Vietnam, whose per capita GDP of $1,910 is less than 4 percent of the U.S. per capita GDP, outscored the United States impressively in math and science.

In math, fewer than 9 percent of U.S. students scored at the advanced level; compare that to 55 percent of students in Shanghai, 40 percent in Singapore, and 16 percent in Canada. An even more disturbing trend is that when American students are in elementary school, they actually compare well to students of the same age from other countries, but as they move to higher grades, they fall behind dramatically.

Even America's highest-achieving students often struggle to maintain their elite performance. A 2011 study found that as time went on, early high achievers failed to improve their reading ability at the same rate as even average or below-average children did. Though there are bright spots (mostly in the wealthiest U.S. schools), the trends are clear: educators are generally failing to cultivate our most promising students as they grow.

All of which suggests that we may be squandering a crucial national resource: our best young minds. "These are the people who are going to figure out the riddles," says Lubinski. "Schizophrenia, cancer, how to fight terrorism; they're going to create the innovations that drive our economy. This is the population who you'd do well to bet on."

18

ATOMIC TRAVEL

Now it's tiffany who drives, north along U.S. Highway 84 toward Los Alamos. Taylor, now sixteen, has convinced his mom to bring him to New Mexico for a few days to hang out with Carl Willis, who has become, as Taylor describes him, "my best nuke friend."

Cocking my ear toward the back seat of the rented SUV, I catch snippets of Taylor and Willis's conversation.

"The idea is to make a gamma-ray laser from stimulated decay of dipositronium."

"I'm thinking about building a portable, beam-on-target neutron source."

"Need some deuterated polyethylene?"

Taylor and Willis have invited me along on their latest "nuclear tourism" junket. The plan is to visit the historic nuke sites in and around Los Alamos, prospect for uranium, and scrounge through the desert in search of the still-radioactive detritus strewn—by plan or by accident—by atomic-weapons developers and deployers.

Taylor and Willis first met in person when Taylor was twelve, when the family stopped in Albuquerque on a cross-country road trip. At that point, Taylor had collected most of the parts for his reactor, and he and Carl were communicating regularly about the finer points of fusor construction.

"I think by then you'd found a source of tungsten wire for the grid and brought it with you," Willis says, "and we were trying to figure out a way to fabricate it."

They were drawn together by their shared passion for nukes, and their age difference quickly became a nonissue. "We always had these great conversations, not just about the technical stuff but about the history and philosophy of nuclear stuff," Taylor says. "I never felt like a normal twelve-year-old around Carl; I felt like a peer."

"Taylor, let's face it," Willis says, laughing, "you never *were* a normal twelve-year-old. Even back then, you knew more about the things I was interested in, more enigmatic nuke stuff, than any PhD or postdoc I knew."

Willis is thirty now; he's tall and thin and much quieter than Taylor. When he's interested in something his face opens up with a blend of amusement and curiosity. When he's uninterested, he slips into the far-off distractedness that's common among the supersmart. Tiffany asks him how his work is going, and Willis says he's thinking about leaving his nuclear engineering job at an Albuquerque company that makes particle accelerators to develop neutron generators at a small R&D company. "Whatever I do, I want to stay in Albuquerque," he says. "It's a good climate; there's lots of labs, lots of uranium, lots of radioactive stuff. And there's lots of history out here that's tangible and collectible —as long as you come with a Geiger counter."

Taylor and Willis typically get together two or three times a year. They scavenge for equipment, visit research facilities, prospect for uranium, or run experiments. As we drive, they talk about taking their next atomic adventure farther afield. Willis says Taylor would love Chernobyl (Willis has visited twice), where "there are still pockets of surprising radioactivity." Also high on both their lists is Shinkolobwe, the mine in Congo that yielded much of the material for the first atomic bombs, and the Semipalatinsk Test Site in Kazakhstan, the highly contaminated Soviet counterpart to the U.S. military's Nevada Test Site. Soviet officials "told the top scientists that they would emerge at the end of a successful nuclear test as national heroes," Willis says, "or, if it failed, they would be shot on the spot. The lower-ranking people were threatened only with prison camp in Siberia."

The U.S. nuclear weapons program, based in Los Alamos, had more of a carrot than a stick approach, appealing to the scientists' sense of patriotism. In late 1942, Enrico Fermi's team in Chicago succeeded in creating the first self-sustaining fission reaction, proving his prediction that nuclei of uranium atoms would split if bombarded with neutrons, releasing more neutrons that in turn would split more atoms, creating an ongoing and self-sustaining reaction. The gates of the universe had been flung open, to paraphrase Pearl Buck, and mankind passed into new galaxies of possibilities that were both majestic and terrible.

The U.S. quietly and quickly accelerated research into nuclear weapons. The Manhattan Project's scientific director was Robert Oppenheimer, a stick-thin, hard-drinking, extraordinarily ambitious man who had been a child science prodigy. Oppenheimer, who had visited the Los Alamos area during childhood family vacations, decided that the remote desert where he and his brother had ridden horses would be the perfect setting for a secret complex of atomic-weapons laboratories.

We pull into town, passing contaminated ponds and gullies, and stop at a diner for lunch. Taylor, wearing his belt-mounted radiation detector, heads for the restroom as Tiffany scans the menu. He returns with news. "I would like to point out that the bathroom is radioactive," he says.

"Did you wash your hands?" Tiffany asks.

Taylor likes to wear his detector in public places to see what—or who—is radioactive. "I'll be in the mall sometimes and it'll go off as I'm passing someone. Usually they're nuclear medicine patients. A lot of them are surprised when I tell them they're radioactive."

"I used to do that too," Carl says. "But now I'm getting old enough that it's not cute anymore. It's just weird."

Taylor eats a third of his lunch, then puts his fork down. "Whew," he says. "I was starving!"

The talk gets back around to Taylor's first fusor attempt, when he was twelve. By the time he loaded his self-made deuterium gas into his fusor, he'd already realized that it wasn't going to work. "I knew I didn't have the vacuum, the voltage, the chamber, or the grid I needed. Looking back, there was just no way I had the technical chops to even come close. But I learned a lot and it was fun to fiddle with, and I think I enjoyed deluding myself that it might work."

Taylor decided to start over from scratch, applying what he'd learned in his first go-round. "Carl gave me a reality check about the voltage and vacuum issues," he says. "He helped me realize that I'd need way more power and that it would be extremely difficult to reach a high vacuum without some ConFlat fittings."

ConFlat, or CF, fittings use knife-edge flanges and a copper gasket to achieve an ultratight seal. Because they're manufactured to exacting tolerances, they can be prohibitively expensive.

"That's one of many reasons why creating nuclear fusion is out of reach for the average person," Willis says. "Usually only the well-financed

An early interest in construction and heavy machinery would be familiar to most parents of young boys. But Taylor's obsessions were extreme, and his parents supported his intellectual passions in ways that few parents would. *Wilson family*

Taylor and Joey with their grandmother Nell, just after she moved into the house next door. Her death from cancer would affect Taylor profoundly, inspiring his vision of himself as a future groundbreaking physicist who would use fusion to create lifesaving medical isotopes. *Wilson family*

Taylor at six and Joey at three. "This might have been Halloween," says his mother, Tiffany, "but this is how Taylor dressed almost every day. And sometimes he could get Joey to follow suit." The boys were close in age and ability, but as they grew, their personalities would evolve in very different directions. *Wilson family*

When Taylor was nine, he and his father, Kenneth, spent three days at Space Camp, a program offered by the U.S. Space and Rocket Center. Talent-development researchers say that early novel experiences spark creativity and innovation and help shape healthy brain systems that enable effective learning. *Wilson family*

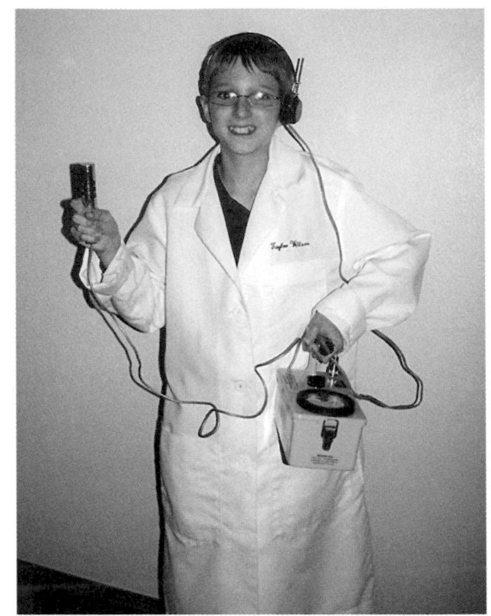

A budding scientist's Christmas: Taylor, in the fifth grade, wears the monogrammed laboratory coat his grandmother gave him and shows off the new Geiger counter that Santa left under the Wilsons' Christmas tree. *Wilson family*

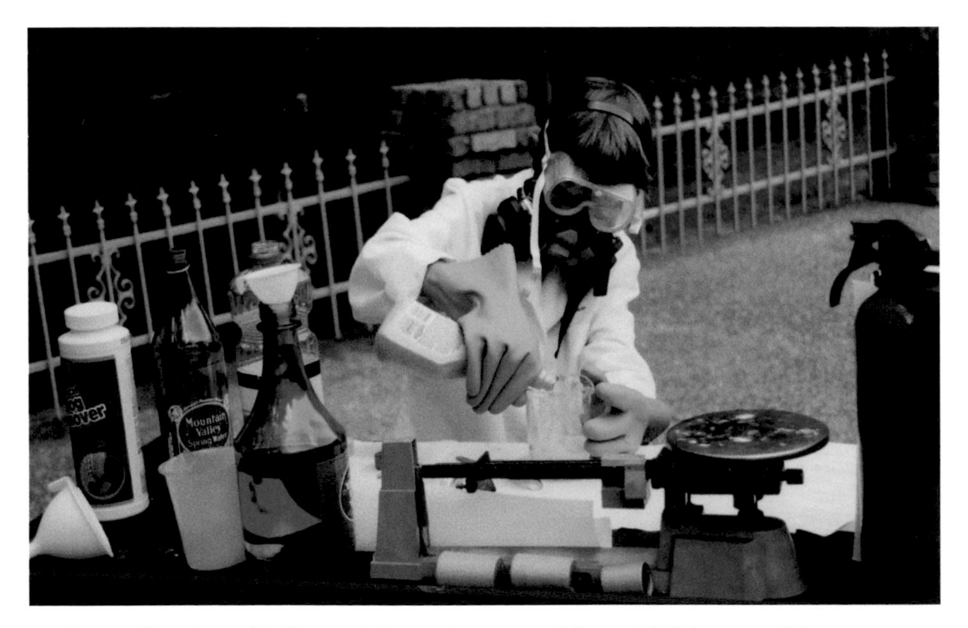

Explosive chemistry: Taylor at eight or nine years old, outside his garage laboratory, mixing methanol and sodium hydroxide as part of his biodiesel production project. He vainly tried to convince his father to use the homebrewed fuel to run his Coca-Cola delivery fleet. *Wilson family*

Taylor with his "best nuke friend," Carl Willis, outside the Black Hole, near the Los Alamos National Laboratory. Now defunct, the unusual surplus store sold precision hardware jettisoned by the weapons industry, which made it a mecca for high-energy science enthusiasts from all over the world. *Tom Clynes*

Taylor in his garage laboratory in Reno. As his collection of radioactive materials grew, so did his parents' worries. *Deanne Fitzmaurice/National Geographic Creative*

With Bill Brinsmead and Ron Phaneuf, in Phaneuf's laboratory at the University of Nevada, Reno, where Taylor built his fusion reactors and other inventions. The two men would emerge as Taylor's primary mentors over four years. "We learned as much or more from him," Phaneuf says, "than he learned from us." *Tom Clynes*

Taylor at fourteen, just after achieving nuclear fusion, with his reactor in the subbasement laboratory at the UNR physics department. The blackboard behind him is covered with calculations and a few holes burned by a laser that Taylor and Bill Brinsmead experimented with.

Photo by Mike Wolterbeek, University of Nevada, Reno

Inside the reaction chamber of Taylor's fusor, as the negatively charged tungsten grid begins to glow. The grid attracts positively charged deuterium ions into a superheated plasma, where some will collide and fuse, releasing energy in the form of neutrons.

Taylor Wilson

Taylor running his fusor's control panel, shielded from the reactor's lethal radiation field by a wall of lead blocks. Safety concerns led Taylor's mentors to take extraordinary measures to protect him. As Ron Phaneuf explains, "We have a lot of radiation sources in our laboratories. Taylor was young and enthusiastic, and enthusiasm doesn't protect you."

Tom Clynes

Physicist and plasma researcher Ron Phaneuf cleared out a corner of his laboratory for Taylor and advised him as he pursued his ambitious projects. Most pioneering and high-achieving adults say that having a mentor was important to their development — but meaningful mentorship opportunities are scarce in today's educational environments. *Deanne Fitzmaurice/National Geographic Creative*

Taylor clowning, Slim Pickens–style, on Bill Brinsmead's replica of the Little Boy bomb — a campy nod to the movie *Dr. Strangelove* — that Brinsmead rides each year at the Burning Man festival. *Tom Clynes*

Nuclear family: After becoming the youngest person on Earth to achieve nuclear fusion, Taylor poses with his family for a photo shoot for *Popular Science* magazine. His early rise to fame would create exceptional opportunities as well as some darker consequences for the entire family. *Bryce Duffy*

Demonstrating the small reactor that Taylor developed for inspecting cargo, at the 2011 International Science and Engineering Fair. "Some [physicists] like working on the questions and the mysteries of the universe. Me, I like applying [physics] to real-world problems, whether it be terrorists bringing nuclear weapons through ports or curing cancer with radioactive material." Taylor would win big at ISEF that year. *Society for Science & the Public*

Taylor speaking at the TED Conference in 2012, at age seventeen. The talk, which TED titled "Yup, I Built a Nuclear Fusion Reactor," was unscripted (as all but one of his presentations have been) and received a standing ovation. The online video of the talk has been viewed more than 2.5 million times. *James Duncan Davidson/TED*

Taylor skipped classes in order to attend the 2012 White House Science Fair, where he and President Obama chatted and joked for several minutes. "You may be working for me soon," the president told Taylor as they parted.

Official White House Photo by Pete Souza

Davidson Academy's graduation ceremony with director Colleen Harsin (left), and Davidson Academy founders Jan and Bob Davidson. As government support for gifted-education programs has dried up, philanthropists such as the Davidsons have stepped in with funding and innovative programming. "We looked at what works best in education and we applied it to gifted kids," says Bob Davidson. "We didn't listen to the people who've done education wrong for a hundred years." *Davidson Academy, photo by Theresa Danna-Douglas*

On the set of *The Big Bang Theory:* Taylor sits in physicist Sheldon Cooper's usual spot, next to actor Johnny Galecki, who plays Sheldon's roommate, Leonard Hofstadter. While on the set Taylor made a discovery that would become a permanent part of the television show's lore. *Tom Clynes*

companies can afford new ConFlat and high-voltage equipment and other precision stuff. The used stuff is in high demand, but if you're savvy and highly motivated, sometimes you can track it down.

"One reliable place to find it," he says, looking at his watch, "is where we're about to visit."

In 1969, an eccentric nuclear technician named Edward "Atomic Ed" Grothus left his job at Los Alamos National Laboratory and turned to selling what he called "nuclear waste"—the lightly used precision hardware jettisoned by America's lavishly funded weapons industry. Grothus informally named his surplus business the Black Hole, since, as he said, "everything goes in, and not much seems to come out."

Grothus hadn't simply quit his job at the weapons lab; he'd stormed out after an epiphany. "What happened here and in Japan in 1945 is a demonstration of the power no person should have," Grothus wrote on a placard that's hung near the front of the store. Renouncing the industry that had long been his—and the town's—bread and butter, he became an antiwar activist. He funded his new life, with exquisite irony, by selling the high-tech jetsam of his former workplace. Grothus also purchased the nearby Grace Lutheran Church building and renamed it the First Church of High Technology; he declared himself cardinal and ministered a Sunday service he dubbed the "critical mass."

We drive past the church and pull into the Black Hole's parking lot, strewn with bomb casings and massive vacuum chambers. Taylor says he found several essential parts for his second fusor attempt among the store's chaotic shelves.

Taylor and Willis leap from the car and pluck their radiation detectors out of the trunk.

"Welcome," Taylor tells me, "to geek heaven."

Geiger counters in hand, the two of them race toward the entrance.

"Mom," Taylor yells over his shoulder, "bring your wallet!"

Taylor and Willis quickly disappear among tall shelves heaped with gas masks, electron microscopes, photomultiplier tubes, detonator cables, and endless piles of esoteric instruments.

Though Ed Grothus died of colon cancer nearly a half decade before the year he'd predicted humanity would blow itself apart (2013), the store is still tottering on when we visit (it would last another six months). Grothus's son, Mike, lives in Los Angeles but has returned on the day of our visit to take care of some business. He tells me that even though the store is becoming less and less viable as a business, he wants to keep it open as long as he can, since "it's a mecca for science geeks from all over the world.

"But we don't get many who take it that far," he says, nodding toward Taylor and Willis, who are prowling an aisle with their beeping Geiger counters. I fall in behind as they home in on something radioactive. When Taylor's clicking slows slightly, Willis, behind him, lowers his probe toward the concrete floor and zeroes in on a brown stain. "Hmm," he says. "I wonder what spilled here."

As Taylor and Willis continue their hunt, Grothus takes Tiffany and me outside and opens a shipping container to show us his father's final big project, a pair of forty-ton white granite obelisks with Grothus's gospel of the atomic threat inscribed in fifteen languages. "These were his Doomsday Stones," Grothus says. "My dad was a character, and by far the most famous person in town. But most people accepted him. It's not as though no one else working in those laboratories had doubts about what they were doing."

One of the wrenching ironies of the American nuclear weapons program is that many of the most creative minds of the twentieth century ended up producing the most destructive force in history. Many pondered the ethical implications of their work; 155 Manhattan Project scientists even signed a petition to President Truman opposing the bomb's use. Oppenheimer delayed the document's delivery until it could no longer make a difference, but after the bombs were dropped, the brilliant physicist's life imploded with guilt. He became a passionate opponent of the escalating madness that gripped the world as the weapons he'd pioneered took on ever-larger dimensions of destructive potential. He opposed the development of the hydrogen bomb with a vehemence so vocal that his security clearance was eventually revoked and he was shut out of Los Alamos, the atomic complex he'd built.

An hour later I locate Tiffany near the front of the store, absent-mindedly poking at her smartphone. "They're lost somewhere in the vacuum section," she says. I find Taylor with a shopping cart half full of equipment. "Look!" he says, lifting up a fistful of gear. "I got this noise-generating tube that's really radioactive and some ConFlat 1.33 connectors."

Nearly another hour later I run into Grothus again and ask if he's seen Taylor. "The kid with the real thick Southern accent?" he says. "He's up front, arguing with his mom."

I approach slowly and spot Taylor and Tiffany near the cash register flanking a cart full of ConFlat fittings, radioactive resistors, and two high-voltage feedthroughs.

"Tay, these high-voltage whatchamacallits are pretty expensive," Tiffany says. "What do you need 'em for?"

"For the fusor I'm thinking about building in the house."

"Tay, no way," Tiffany says.

"I'm thinking just a small one, a little side project."

Tiffany, who had been pulling her wallet from her purse, stops and spins toward her son.

"Taylor," she says, "you are *not* going to build a nuclear reactor in the house."

Later, we'll explore Bayo Canyon, the eerie site of the Manhattan Project's Technical Area 10. Here, engineers tested the implosion design for what became the Fat Man bomb, staging explosions that intentionally spewed radioactive lanthanum (RaLa) into the air—in other words, setting off dirty bombs—and measuring the gamma rays. The RaLa tests were the single most important experiments affecting the final design of the bomb that leveled Nagasaki, which may partly explain the eerie atmosphere in the canyon. We don't talk much as we work our way across the scrubland with Geiger counters and metal detectors, finding bits of still-radioactive debris and skirting fenced-off areas that are still too radioactive, decades after the tests ended, to enter. At one point I stub my toe on something hard and look down at a slab of concrete poured over a stretch of ground. It's embedded with a bronze marker warning *Do not disturb until 2142.*

We linger a little too long and find ourselves locked in at the gate. After a while a guard comes along and releases us, and we head back toward Albuquerque. On the way, Taylor and Willis reminisce about Taylor's second attempt at building a reactor. Taylor, getting more savvy, began monitoring several websites each day, watching for new postings of used laboratory equipment. Surfing the web with his list in hand, he learned to pounce quickly if he saw something he needed at a reasonably good price. He also began contacting anyone who might have a part or a lead for him. One day, he called Theodore Gray, the science author, app developer, and obsessive collector of elements. (Among his most famous projects is a four-legged physical table assembled from thousands of samples of elements from the periodic table.) Gray, an advocate for exposing children to the wonders of science, was glad to talk with Taylor about their respective collections. When Taylor asked Gray if he happened to have any extra tungsten lying around, Gray said he did indeed have a bit of thin tungsten wire, which he'd be glad to send to Taylor for his fusor's grid.

Then, in the final days of the summer before he began eighth grade, Taylor came across something even more promising: a former astronaut living in Houston wanted to sell a mass spectrometer, a large machine that measures the masses and relative concentrations of atoms and molecules. It wasn't fully functional, but the parts Taylor wanted were in good shape. Taylor doesn't remember the price or the astronaut's name, but after they talked for a while, Taylor asked if he'd consider donating it for Taylor's fusor project.

"Imagine," says Willis, "you're an established scientist and you get a call from a preteen kid who wants you to part with an expensive precision instrument so he can use it for the nuclear reactor he's building. There's both absurdity and charm to it, but that's the way it works with Taylor—doors just open."

Kenneth had always been there to enable and support Taylor when he needed materials, mentors, or experiences to feed his evolving interests. Now, Taylor was taking the lead, reaching out to build his own relationships and shape his own trajectory. In the case of the mass spectrometer, though, Taylor would need Kenneth's help.

The astronaut said he'd donate the machine to Taylor if he could figure out how to get it from Houston to Texarkana. As it turned out, Kenneth's company had just begun sourcing cans from the Houston area. The truck driver made a detour to the astronaut's house on his next trip and loaded up six pallets of equipment using a rented forklift.

"I had to hand it to Taylor," Kenneth says. "He did pretty well on that one."

Taylor took apart the mass spectrometer and cannibalized its ConFlat fittings and other components. Though he didn't have a precision workshop or tools or a budget for high-end equipment, he had a strategy that, he was convinced, could work this time: he'd create a more realistic and coherent design, put his money and energy into getting better parts, and resist the temptation to start building the reactor until he'd collected the key pieces of equipment he needed.

Taylor used his time during science lab—"which was about as useless as you'd think"—to diagram and rediagram the fusor's design. At night he'd get online and work the auctions and message boards; he snapped up a thick-walled reaction chamber, a high-voltage power source, and a vacuum pump that was twice as powerful as his last one.

As Christmas approached, he'd acquired most of the fusor's parts and was looking forward to assembling and testing the machine. He figured he'd again need to use his homemade electrolysis gas generator to convert the deuterium fuel from heavy water. But to his surprise, he found a cylinder of deuterium under the tree on Christmas morning, a red ribbon tied around its valve fitting.

"Remember that, Mom?" Taylor says as Tiffany drives. "You guys finally decided I could handle it, and I was so happy!"

"My parents would have never done that," Willis says enviously, shaking his head. He grew up in Oak Ridge, Tennessee. With all its secret histories and its ongoing research, Oak Ridge was, even in the 1980s, an exciting place, Willis says, "for a young science nerd."

"I remember when some plutonium dropped onto the railroad tracks, and another time when my neighborhood was locked down due to a leak of iodine-131."

Willis tells us that his dad was a physicist, his mom a geology teacher.

"I would have loved to have had scientist parents," Taylor says.

"Hey!" Tiffany says. "You got those genes from somebody."

"Keep in mind, Taylor," Willis says, "that my folks were scientists, but they weren't very supportive of the kinds of things I wanted to do. In fact, they were totally unpermissive about radioactive materials of any sort."

Ironically, Tiffany and Kenneth's lack of expertise may have been helpful to Taylor's development as a scientist. Lacking such realistic guidance, Taylor never learned that he couldn't do things; instead, he learned *how* to do things.

Willis's parents supported his early interests in minerals and fossils, but when he got interested in nukes, he says, "I had to be clandestine."

Once free of his parents' watchful eyes, Willis was like a college freshman who'd been denied candy at home gorging on Snickers bars in his dorm room. When we arrive at his house, on Albuquerque's south side, Taylor's belt-mounted radiation detector starts beeping furiously the moment Willis swings open the door.

"Maybe," Willis says, "you could turn that off so you don't drive us all crazy."

Willis, who lives with another nuclear engineer, says break-ins are common on his block. "But I'd pity the poor guy who got in here and slipped the wrong thing into his pocket."

As he takes us through his quack-cures collection in his dining room, he enters an ironic, wry-and-dry showman mode: Stephen Colbert meets P. T. Barnum.

"Ah, the good old days," Willis says, picking up a postcard with a dollop of glow-in-the-dark radium paint, "when you could send loose radioactive contamination through the freaking mail!" He holds a Geiger counter up to the card. "Check it out; it's still putting out eleven thousand CPM!" (counts per minute).

"And look at this! Back in 2004, an average Joe could buy uranium oxide from MV Laboratories in New Jersey with nary a question asked." He hands me a sealed thirty-gram bottle of greenish-black uranium-308. "Now, forget it. Another emblem of American freedom eroded by the drumbeat of irrational fear . . .

"Fortunately, we can all still own some antimatter," he says, grabbing a shaker of potassium chloride salt substitute. His detector clicks as he explains that potassium-40 has a positron decay channel; when an emitted positron and an electron collide, they annihilate to produce two photons.

We follow Willis into his bedroom, where his fusor rests next to a chunk of bomb slag from the Trinity nuclear test and dozens of other gadgets and artifacts.

Tiffany looks worried. "Carl," she says, "I'm a little concerned about you sleeping with this stuff."

"It's all about acceptable risk," Willis says. "Different people would put it in a different place." He looks around the room, spots one of his prizes—a still-functional antique portable x-ray chamber—and snatches it up. "And now," he says, dashing back toward the living room, "how about a classic physics demonstration of the penetrating quality of x-rays!"

Willis and Taylor decide that the coolest handy item to x-ray is a boxed set of socket wrenches. It's interesting to see the tools inside in real time on the screen, but without any shielding I'm starting to wish I'd worn my lead-lined boxer shorts. I begin backing away, as does Tiffany; she discreetly grabs the collar of Taylor's shirt and pulls him along with us.

"Uh, maybe just a few seconds is enough, Carl," Taylor says. Willis nods and flips the machine's switch off.

"It kind of scares me being in Carl's house," Tiffany says as we drive back to our downtown hotel.

"You have to admit I'm better than Carl about that stuff," Taylor says.

"You're not sleeping with it," Tiffany says.

19

CHAMPIONS FOR THE GIFTED

BY THE EIGHTH GRADE, Taylor was spending most of his school days on autopilot, going through work he'd long outgrown as his mind wandered away from classes that were unsuited to his interests, knowledge level, and learning style. Joey, too, now in the fifth grade, was withdrawing emotionally from school, inwardly seething as his math teacher handed out yet another drill-and-kill long-division worksheet.

But Taylor had, at least, made peace with his science teacher. "We were butting heads at first," says Taylor, "but then she just gave up and she let me start teaching." In fact, both Taylor and his friend Ellen Orr often found themselves teaching the classes. "We were in different sections," Ellen remembers, "and we both independently started out helping everyone understand what was going on after we finished our work. Then it got to the point that she just let us take over most days."

"She liked it," says Taylor, " 'cause she didn't have to prep lessons. But I had to prep, so I started missing other classes. It wasn't affecting me academically, but the teachers started to wonder why I wasn't there."

Advocates for gifted children are appalled when they hear stories like this. "It shouldn't be the kid's job, obviously," says Ellen Winner. "But it does point out what some of them have to go through."

At first, Taylor was optimistic about his chances with the second fusor. With the salvaged ConFlat fittings, he got significantly better sealing, but when he began to put it together, it was obvious that the vacuum pump he'd purchased wasn't powerful enough to thoroughly clear the air inside the reaction chamber. He and Willis ran some calculations together and determined that the power source was probably not going to cut it either. What's more, Willis said, without a view port on the chamber, there would be no way to see what was going on once he fired it up.

"So I'd be running blind," Taylor tells me, the echoes of frustration coming through in his voice. "The grid could be melting and I'd never even know it." What's more, the grid itself was primitive. Though

Taylor had convinced Theodore Gray to give him some nice tungsten wire, there wasn't a nearby machine shop capable of fabricating a hub that could secure the wire pieces in a precise spherical shape that could effectively contain a plasma field. Taylor tried to build it by hand, but after a few tries, he set it aside. The project, it seemed, was going nowhere, just like his education.

"Seventh grade wasn't bad," he says, "but by the eighth grade I was done in. I was bored, they wouldn't let me do the things I wanted to do. It felt like each school day lasted a hundred hours. I was always just waiting."

Waiting was the most common response when Tracy Cross of the College of William and Mary asked thirteen thousand kids in seven states to describe in one word their experience as gifted children. "They said they were always waiting for teachers to move ahead, waiting for classmates to catch up, waiting to learn something new—always waiting."

Taylor had dreamed big. His audacious vision of himself creating nuclear fusion and cancer-catching medical isotopes had fueled him and propelled him forward this far. But now his dreams—the reactor he wanted to build, the neutrons he wanted to generate, the pioneering nuclear physicist he would become—were starting to seem like they'd always be just dreams. Taylor had wanted to build his own star, literally. But he'd need more than his own vision and ambition to accelerate particles at speeds and temperatures high enough to fuse atoms. To go back to an earlier question, how could a middle-school kid living on the Texas-Arkansas border ever hope to do that?

The reality is that without some serious support—daily hands-on expert guidance, precision equipment and tools, a well-equipped laboratory—he probably couldn't. And that, say Janice and Robert Davidson, illustrates the biggest tragedy in American education.

Though Taylor the eighth-grader had never heard of Jan and Bob Davidson, they would soon become two of the most significant figures in his life.

Software multimillionaires who developed the Math Blaster and Reading Blaster educational software in the early 1980s, the Davidsons have been championing the idea that the most underserved students in the nation are those, like Taylor, who have big brains and big ambitions.

"There are so many kids like him, whose talents are wasted in unchallenging environments," says Bob Davidson.

Bob, in his seventies, talks fast and decisively and interjects what seems like a well-rehearsed touch of outrage when he talks about his favorite issue, though at other times he can be quick with a joke or a comic comment. He's tall and he moves quickly, in all aspects of his life. In addition to his fast gait, he races sports cars. (In 2014, he drove over the finish line as his team took first place at 25 Hours of Thunderhill, the world's longest endurance road race.) As a young man, he burned through degrees in chemical engineering, business, and law before working his way up to executive vice president of an engineering company. Then he came on as CEO of the educational software company his wife had founded.

Jan, a year younger, is blond and petite. She earned a PhD in American Studies, then, in the 1970s, became a computer pioneer after she had an epiphany that computers had the potential to individualize instruction and make learning fun. Math Blaster debuted in 1983 and was an instant hit, using games to teach and reinforce mathematical concepts. The Blaster Learning System expanded to other subjects, such as reading and science, and in 1996 the Davidsons purchased Futurekids, a pioneering computer and technology training program that focused on tutored computer education centers and preschool technology literacy.

The Davidsons sold their original company in 1997. Looking for a way to put their fortune and experience to work to influence school reform, they spent more than two years reviewing education research. In 1999, the answer came when Bob read a newspaper story at their retirement home in Lake Tahoe. It was about a five-year-old in Maryland who was reading at an eleventh-grade level, but the schools said he had to start school in kindergarten. "Imagine that!" he says, outrage creeping into his voice. "You've got this genius kid and these parents trying to advocate for him so he can reach his potential, and the schools are basically saying the bureaucracy is more important than the child."

The couple decided that education for profoundly gifted children would be their target mission, reasoning that the population was small enough that they could make a significant impact. In 1999 the Davidsons launched a Young Scholars program that now provides mentorship and advocacy to about three thousand U.S. scholars who qualify with standardized test scores in the 99.9th percentile or IQs of at least 145. They followed that with the Davidson Fellows, a sort of junior MacArthur grant for children, and the Think Summer Institute, which offers college courses for twelve- to fifteen-year-olds. They've also funded educational research and cowritten a book (with Laura Vanderkam) titled *Genius Denied: How to Stop Wasting Our Brightest Young Minds.* In 2006 they took their most audacious step yet when they opened Davidson Academy on the University of Nevada–Reno (UNR) campus. The school was a response to ongoing suggestions from parents of children in other Davidson programs that there was a need for a bricks-and-mortar school for their children so they could learn at an accelerated pace in the company of their intellectual peers and pursue the subjects they were passionate about. The academy was established as a subsidized, tuition-free public school for the profoundly gifted.

Reno, not known for intellectual vigor, seems an unlikely location for a school for geniuses. But Jan says that when they surveyed parents of potential students, the consensus was "Wherever you put it, we'll come." The couple had political connections in Nevada, people who helped them get state education laws changed to allow their experimental school to teach almost any way it wanted, though the academy has to meet core curriculum requirements and participate in state testing.

Jan and Bob also had connections with members of the University of Nevada–Reno's leadership, who were open to having the academy on the campus. The couple spent $3.7 million converting the university's former student union into a suitable space for their school, which accepted thirty-five students its first year; there are about 140 now.

Students above the IQ or standardized-test threshold are invited to apply to the academy. Applicants then go through a demanding all-day assessment that weeds out about two-thirds of the candidates. About a third of the students relocated from out of state with their families to attend the academy.

There are no quotas for ethnicity, financial need, or anything else; if the child demonstrates ability and the family is willing and able to come to Reno, the child is in. The school is, in many ways, a reaction to trends favoring equity over excellence, a subject that Bob will argue passionately about.

"As a society, we are selectively elitist and anti-elitist. Certain types of childhood giftedness—such as athletic talent—are supported without question. But grouping and accelerating by intellectual ability is less palatable than grouping by athletic ability." Parents understand that only a small fraction of young athletes will become professionals. "But with academics, achievement leads to success and privilege, and accelerated learning assists some people to become part of the privileged class. Parents are worried their children will be left behind, so they put equality over achievement. We've got to get beyond that. Equality is giving every child an opportunity to excel, not making everyone have the same percentage at hitting a free throw."

Tiffany and Kenneth found out about Davidson Academy after Kenneth's daughter, Ashlee, came across a 2007 article about it in *Time* magazine and sent a copy to Texarkana. It looked interesting, but moving across the country—leaving behind family, friends, and two businesses—wasn't something Tiffany and Kenneth wanted to do.

"We have lots of roots in Arkansas, and we knew nothing about Reno," Tiffany says. "But we could see the potential of this type of learning environment. Taylor was headed toward high school and he wanted to pursue his passion in nuclear science. And Joey was about to start sixth grade and we needed to find a more challenging school for him. We started to seriously consider it."

Tiffany and Kenneth decided to bring the boys to Reno, take a look around, and have Taylor and Joey tested for admission. But they agreed in advance that they'd consider going only if both boys were accepted.

"We actually thought the most likely scenario was that Joey would get in but not Taylor," says Tiffany, since Joey had always scored higher than Taylor on every standardized exam, from his kindergarten-readiness test to his ACT.

But Joey wasn't at all keen on moving. Even though he was miserable in school, he didn't want to leave his friends behind. Taylor, however, loved the idea of hanging out with other science-focused kids and having access to courses and maybe even laboratories in the University of Nevada's physics department, one of the strongest departments in what is not a particularly prestigious university. Taylor got even more excited when he found out that the department's professors included Friedwardt Winterberg, a German-born theoretical physicist who had studied under Nobel Prize–winning quantum physicist Werner Heisenberg. Winterberg, whose main interest was nuclear fusion, had come to the U.S. through Operation Paperclip, which scooped up German scientists after the war. A noted theoretician, Winterberg had developed a concept for a nuclear fusion propulsion reactor for space travel.

"When I read about that, I knew I wanted to go there," Taylor says. "I couldn't wait to meet him!"

PART IV

20

A HOGWARTS FOR GENIUSES

ONCE THEY GOT WORD that both Taylor and Joey had been accepted to Davidson Academy, Tiffany and Kenneth considered the pros and cons and decided that it was worth the sacrifices they'd have to make to give their boys what seemed to be an outstanding educational opportunity. Kenneth asked Tiffany's brother if he'd be willing to step in to run the Coke company, and Tiffany asked a friend if she'd like to take over for a while at her yoga studio and juice bar. Both said yes.

"We thought maybe we could just do it as an experiment, an adventure for a year or two, then we'd come back," Tiffany says.

Two years later, Taylor and I drive north on Reno's Virginia Street, passing under the arch that has long welcomed visitors to THE BIGGEST LITTLE CITY IN THE WORLD. Taylor, on the passenger side, flicks the hair out of his eyes and looks around as we continue up the gaudy strip, going by the casinos and through downtown and the porn district, past tattoo and piercing parlors.

"At first we'd laugh when we drove through here. Joey and I would say, 'Isn't it funny that kids should have to see this stuff on the way to school?' But I don't mind; it's just part of life here."

Taylor is looking forward to showing me around Davidson Academy and, especially, to introducing me to Elizabeth Walenta, his chemistry and physics teacher. "Actually, Ms. Walenta is kind of more of a colleague than a teacher," Taylor tells me after we park and begin walking toward the UNR campus. "I come in and teach the classes sometimes, and recently I vetted the particle physics course at the university for her; I gave it the thumbs-up."

Apart from the fingerprint scanners at the door, Davidson Academy looks a lot like a typical high school or middle school, and the kids could easily be mistaken for suburban American students anywhere —until they open their mouths. It's then that you realize that this is an exceptional place, a sort of Hogwarts for brainiacs. As these math whizzes, musical prodigies, and chess masters pass in the hallway, the banter flies in witty bursts, and you rarely hear the word *like* used more than once in a sentence. Inside humanities classrooms, discussions spin into sharp intellectual duels.

"I thought maybe everyone would be nerdy, with no social skills," Taylor says as we walk through the hallway. "But only maybe ten or twenty percent are like that. It's been fantastic, socially."

Then again, it could be better. In the hallway, we run into Sofia Baig, an athletic, dark-haired girl on whom Taylor has a very public crush. Unfortunately, she has a very public, very steady boyfriend.

"Hey!" Sofia says as she walks by, flashing a broad smile and playfully punching Taylor's shoulder.

Taylor turns and watches her walk away.

"Hey!" he calls after her. "You're awesome!"

While Taylor attends his Critical Theory class, the school's director, Colleen Harsin, shows me around, peeking into classes, stopping to talk with faculty and students. "These are kids who wanted to go above and beyond the standard classroom curriculum, but their schools lacked the resources or the motivation to let them. You'll notice that classes move very fast here, and the kids demand that—so it takes a special kind of teacher.

"Everyone has some sort of advanced obsession," says Harsin, who wears bright print dresses and has the open but no-nonsense demeanor that tends to work well for principals in the upper grades. "Taylor is Mr. Nuclear, that's his thing. We've got a ten-year-old in calculus, and a girl who just won the International Informatics Competition."

These are the kind of kids that a typical teacher might encounter once in an entire career. Their scores on standardized tests (the SAT, the ACT, or the EXPLORE) are in the top one-tenth of 1 percent. In most classes, children are working at least three grade levels above what they'd normally encounter at their age, and students are grouped according to ability rather than age. The main goal, Harsin says, is to be flexible and supportive so that students are not held back in their areas of passion. "If one of our students is ready to have space in a university laboratory and run grant-based experiments," says Harsin, "they can do that."

Small class sizes allow instruction to be tailored to the students' specific interests, with the goal of going deeper and integrating the subject matter into long-term pursuits. On Fridays, students explore unfamiliar subjects in elective courses like Astronomy, Neurophilosophy, and even Vampire Literature. "Considering that everyone's so focused, that's been really interesting, to watch them expand their intellectual horizons," says Harsin. "I've seen a few discover a whole new passion and run with it."

The academy, which accepts students as young as nine, employs UNR honor students to accompany children under sixteen to classes at the university.

"We used to call them escorts," Harsin tells me. "But somebody suggested that, considering we're in Reno, it might be better to call them chaperones. I immediately said yes to that idea."

There's currently no other public school doing things quite this way, although some expensive private schools are similar. Most Davidson families are middle-class, but according to Harsin, "We have a few students who would qualify for free-lunch programs in other places."

For a teacher, a job at Davidson is a plum assignment. Not because the job pays particularly well (teachers here make the same scale as other Reno teachers but aren't union members and don't get tenure), but because the chance to teach a whole roomful of kids like these—so eager to learn and to explore—comes along so rarely.

"It's not the intelligence," says English teacher Alanna Simmons, "it's the energy and the curiosity, it's where I can take them and they can take me and the chance to go to a place even bigger together. Call it a Utopian approach, but it actually happens here, every day. These kids are so smart and funny and cut through all the noise and bullshit that exists; it's invigorating."

Humanities classes tend to be less age diverse than STEM classes, since ability in that arena is dependent on emotional trajectory and range of experience. In Taylor's first year at the academy, he showed up in Simmons's creative writing class. "He's not what you'd think," Simmons says. "People assume that he's phoning it in in English, but it's just the opposite; he's energetic and passionate. And as a scientist, with critical theory he can come in and look at a work and analyze it and form a critique based on the knowledge-making framework of science. I hadn't seen that before; it's really original. Oh, and he loves to perform!"

Christopher, a student hanging out in Simmons's classroom, chimes in. "I love it when he does his Southern preacher voice. When you watch him, you say he should be an actor."

The long padded benches outside Harsin's office are the primary hangout area. "I call it Shenanigan Central," she says, watching her students clowning. "Some students struggled, socially, at their previous schools. Being around peers that they can relate to and making friends who think as quickly as they do is what most of them say is the best thing about Davidson. They're used to being the smartest by far where they came from; some would formerly even dumb themselves down."

"I never felt like I had to dumb myself down, but you hear of kids having to do that," says Sofia Baig. She and two of Taylor's other friends are hanging out in Shenanigan Central with me, talking about Davidson and their former schools. "My parents came from India, and they were pretty aggressive at getting me into accelerated programs, summer programs, and things like that." Sofia plays softball at the local high school, since there are no sports teams at Davidson. "I do notice, though, that when I'm with students from outside the school, in softball or something, I might use a word and I'll see blank faces."

"At my old place, we never used higher concepts; they just never came up in conversation," says Anthony "Biff" Fidaleo, another student. "Here, we don't have to be conventional; we can study esoteric subjects and get in as deep as we want to."

"There are a lot of savants here," says Tristan Rasmussen. "Biff has his math, and Taylor is obsessed with nukes. In fact, I'm completely convinced that Taylor is going to either win at least one Nobel Prize or be on the *America's Most Wanted* list by the time he's eighteen. Myself, I'm more of a polymath. I was actually functioning well at my old school, enjoying myself. But it was a breath of fresh air coming here. I can take anything I want at the university, so I have thousands of courses to choose from. The kids are interesting here. It's a faster pace."

Biff, who has already maxed out on the highest math class at the university, is in a choir with Taylor and has taken biology classes at UNR with him. "Taylor is the most energetic person I know," he says. "He loves discussion, he loves drama."

Sofia laughs. "He'll come in on a Monday morning all breathless and say, 'I almost died this weekend!' "

Though humanities instruction is strong at Davidson, the academy tends to draw mostly STEM-focused kids. In contrast to most high schools, the kids who are *not* science nerds are the quirky minority. Davidson students say that by and large, the humanities classes at the university have proven to be disappointing.

"In the history classes at UNR," says Tristan, "the students don't discuss; they will sit there. You may get more complex concepts, but the students are not inquisitive and it's not much fun. When we do history here, it's an exploration, it can be wild; everyone's got an opinion and a question. The discussions are explosive."

Taylor comes out of his class looking fired up. "You missed a wild one!" he tells his friends. "*Ulysses!*"

The bell rings. "Okay, off to study hall," Tristan tells Biff, and the two trade high-fives, mocking the brainiac stereotype for my benefit. Taylor leads me through the halls and into the science lab.

"Tom, meet Elizabeth Walenta," he says. "I'm her star student."

Walenta looks up from her desk. "Actually, Taylor's more of a teacher than a student at this point," she says, not missing a beat. "He was my guest lecturer last week, talking about the structure of the atom."

"So, how are the kids doing on the periodic table?" Taylor asks, beginning to rifle through the room's cabinets.

"Really well," she says. "How's that molecular biology class at UNR?"

"I'd give it a seven out of ten," Taylor says.

He finds a Crookes tube in a cabinet and hands it to me, telling me that it's part of his personal collection that's "sort of on loan" to the school.

"Careful!" he says. "That's a hundred years old. J. J. Thomson used one of those to discover electrons. Check it out; it still has vacuum integrity.

"Anyway," Taylor says, continuing to open and close drawers and cabinets, "y'all talk."

Walenta, who wears her hair in a long, no-nonsense braid, came to Reno from Irving, Texas. "With the kids here I don't have to do as much repetition or modeling. I give them the basics, see if they can figure it out, and then they just take off."

Then again, she says, she misses the freedom to do more hands-on science. "Maybe it was a Texas thing," she says. "You want to go and blow something up over in the field, you go and do it. We'd let kids make catapults and have them do their calculations, figure out angles, trajectory, wind speed, then use us teachers as their targets.

"Some kids wanted to make dynamite; maybe that's why I hit it off with Taylor. It's harder here to find places to experiment Texas-style. We did try thermite, which created divots by the duck pond. UNR got mad at us."

Taylor opens a cupboard door marked *Explosive*. "This is always a good candidate to look in," he says, adding, "Remember when I synthesized some nitrogen triiodide and put a little bit on the doors?" Walenta gives him a how-could-I-forget look.

"It's known to science as the most unstable of explosive compounds," Taylor tells me. "A feather waved over it or a fly landing on it can detonate it. I wreaked some havoc with it," he says. "But Ms. Walenta was pregnant, so in retrospect it probably wasn't a good idea."

Walenta laughs. Later, when I stop by on my own, she will tell me: "You know how they say if you want to be happy, be around happy people? It's the same with passion and excitement too. Taylor gets everyone excited in class; he gets everyone into it."

"This shouldn't be here," Taylor says, pulling out a plastic container. "This should be in the poison cabinet."

"You could make an argument for that," Walenta says.

Then Taylor and Walenta fall into a deep discussion about ununoctium, the temporary name for the element with the atomic number 118 discovered in 2002, and the possibilities of a yet-to-be-discovered "island of stability" among the heavier isotopes of transuranium elements.

"If its periodicity is correct," Taylor says, "then ununoctium should be chemically stable, just like its noble gas cousins."

I glance at the clock; any moment now I'll literally be saved by the bell. But just before it rings, a smaller, darker-haired version of Taylor walks through the door. A broad smile comes across Taylor's face.

"Joey!" he yells, walking over to give his brother a big hug.

There's no question that Taylor is thriving at Davidson, socially and intellectually. But for his younger brother, the experience has been mixed. Academically, Joey was stimulated, especially at first. He'd quickly moved through Davidson's math offerings and then into the higher-level courses at UNR: calculus 2, differential equations, statistics, linear algebra.

But Joey's interests began to broaden beyond those of many of his fellow students who'd been drawn to Davidson for the chance to pursue highly specialized endeavors. In his second year, he got curious about Japanese. The mental skills needed for foreign-language learning and math are closely related; those who show talent in one subject are usually good in the other. Joey persuaded two friends to attend Japanese 101 at UNR with him. It was just a matter of clearing it with one of the academy's curriculum directors.

Though his friends were approved to take Japanese at UNR, the answer for Joey was no.

"Probably the only thing worse than a kid being forced into making a choice between friends and academics," says Winner, "is a kid being told that he can't have either."

The issue, apparently, was Joey's performance in English. He was a good writer, his later teachers would say, but a slow one. "It took me a while to even realize it was a problem," he says. "But after a while I could tell the English teachers were getting impatient; they'd come over and stand over my shoulder and I'd start getting nervous that I wasn't doing it right and just delete everything. I guess there was an incompatibility between us. Pretty soon I started falling more and more behind."

On the family's first trip to Reno, even before Taylor and Joey were accepted to the academy, Taylor made arrangements to meet with Friedwardt Winterberg, the UNR physicist who had come to the U.S. in 1959 as part of Operation Paperclip. A protégé of both Heisenberg and Wernher von Braun, Winterberg was a sort of *Dr. Strangelove* figure, brought over in the postwar years so that the U.S. could take advantage of his physics expertise. With thin, receding gray hair and wire-framed glasses, he looks (and is, at eighty-five) substantially older than the rest of the department's professors.

Both Tiffany and Kenneth came to the meeting with Taylor. Winterberg was polite as he invited them into his office and asked them to sit, but he wasn't really sure what was going on: Why did this child and his parents want to see him? He looked at the adults expectantly, but it was Taylor who started talking. This wasn't something the German was accustomed to, but Winterberg listened, making a halfhearted attempt to mask his impatience—right up until Taylor told him that he wanted to build a nuclear fusion reactor.

"You're thirteen years old!" the professor erupted in an accent fairly bursting with umlauts, "and you want to play with tens of thousands of electron volts and neutrons and deadly x-rays?" Such a project would be far too technically challenging and hazardous, Winterberg insisted, even for most advanced doctoral candidates. "First you must master calculus, the language of science," he boomed. "Then work your way up through the theoretical physics courses."

Later, Winterberg would seem nearly as impatient with me when I asked him about the encounter. "He was only thirteen! And he wanted to do nuclear fusion. It was more than a little outrageous, who wouldn't think so? And so I told him to concentrate on his mathematical skills, then move into theory."

Taylor hadn't known that Winterberg was considered by the academic community a brilliant but notoriously cranky naysayer; he'd once suggested in an article that Einstein had plagiarized part of the theory of relativity from the notebooks of David Hilbert.

"He's the resident crazy German that every physics department apparently thinks they need," Taylor would tell me later.

In any case, Winterberg seemed to have put the kibosh on Taylor's nuclear ambitions. "After that," Tiffany said, "we didn't think it would go anywhere. Kenneth and I were secretly a bit relieved."

But Taylor still hadn't learned the word *can't*.

"When Taylor wants something," says Kenneth, "there's no way to stop him. Wherever you try to cut him off, he figures out a different angle."

In this case he didn't need to go far to get around the obstacle. Next door to Winterberg's office, atomic physicist and plasma researcher Ron Phaneuf had an entirely different reaction to Taylor's plans. Taylor went to talk to him shortly after he started at Davidson Academy.

"I was immediately impressed because he had a depth of understanding that I'd never seen in a kid this young," Phaneuf says. "And he was ambitious; he wanted to build a fusor. This isn't something you hear every day. He seemed to have a good understanding of the risks, but he was telling me he had collected isotopes and was proposing to build the reactor at home in his garage. And I'm thinking, *Oh my Lord, we can't let him do that there. But maybe we can help him try to do it here.*"

Phaneuf, who was four years away from retirement when he met Taylor, is thin, with rust-colored hair and a perennial tan. He wears cardigan sweaters and walks easily in his sneakers. When he's expressing something thoughtful, his eyebrows come together, giving him an almost-but-not-quite-worried look.

Phaneuf told Taylor that he'd think about some options.

"And I did—a lot. I knew right away that Taylor was not like anyone I'd ever met before. He's so sincere and forthcoming, and so excited about science that you can't ignore him." Nonetheless, Taylor was very young (fourteen at that point), and Phaneuf could understand Winterberg's reaction. "But he's a theorist; he's old-school. And I think he may have thought Taylor was trying to achieve break-even energy with his fusor. I'm not sure he understood that Taylor's intention was primarily to produce neutrons."

"Winterberg was the department's big name, so it was understandable that Taylor would start with him," says science teacher George Ochs. "But he might have thought Taylor had been spoon-fed information and was just regurgitating it. If he'd taken an interest and gone deeper, like Ron did, he would have found out that wasn't the case. Taylor is nothing like a lab rat; he's a lab shark. He's always going to be the principal investigator on everything he does."

Phaneuf invited Kenneth and Tiffany to stop by his office a few days later. "I could see during that first visit that his parents wanted to nurture his special gift, but they were afraid of it too," Phaneuf says. "I told Kenneth I could help Taylor try to realize his goal. But I also said that the kinds of things he was proposing to do really should be done in a lab, not a garage. When I offered to see if I could find him lab space in the physics department, where we could do it safely, I could see Kenneth's shoulders settling. He was relieved."

Now Phaneuf had to sell the idea of a fourteen-year-old boy working with hazardous materials, high voltages, and radiation to the rest of the physics department. It didn't hurt that at the time Phaneuf was chair of the university's radiation safety committee. He set up a meeting with department chair Roberto Mancini and Davidson Academy officials.

"We were breaking new ground here," says Phaneuf, "bending the rules a bit. Everyone of course had concerns about safety, and the academy wanted to support his project but they were a little afraid, justifiably so. We have a lot of radiation sources in our laboratories. Taylor was young and enthusiastic, and enthusiasm doesn't protect you."

In the end, the officials said that if Phaneuf was confident that it could be done safely, and if Phaneuf could include the project under the university's broad-scope license for radiation-emitting devices, they'd support it. For safety reasons, someone from the department would have to be with Taylor at all times. Young people can be more sensitive to radiation exposure than adults; their smaller bodies absorb more radiation and their cells are more vulnerable because they're multiplying quickly.

The next time Phaneuf and Taylor met, the professor invited Taylor to sit in on his upper-division nuclear physics class at the university. Then he offered to clear out a corner of his own research laboratory for Taylor's project. "You should have seen his face light up!" Phaneuf says.

Phaneuf then approached some of the other department physicists and technicians. Most thought Taylor's chances for success were pretty low, but they were intrigued by this fourteen-year-old nuclear physicist. They said they'd be willing to try to accommodate him.

Phaneuf introduced Taylor to a colleague, computational and plasma physicist Bruno Bauer, whose daughter Rebecca was a classmate of Taylor's. Taylor asked Bauer if he had any broken pieces that Taylor could use for the fusor. Bauer said they might be able to do better than that and suggested to Phaneuf that he bring Taylor down to the subbasement.

"That's where they keep all the dangerous stuff," Taylor says. "And that's where I met Bill."

If you've been to the Burning Man festival, you've likely seen Bill Brinsmead; he's the guy wearing a tie-dyed bathrobe riding atop a wheeled replica of the Little Boy bomb — a campy nod to *Dr. Strangelove.*

"I was afraid that people might see it as a kooky glorification of war," says Brinsmead. "But they all get it: It's a fat man riding Little Boy."

Taylor and I meet up with Brinsmead at the Hub, a coffee shop where Brinsmead holds court each morning. Brinsmead has a shaggy mustache and solidly muscled arms and wears a gold chain with a Buddha encased in a plastic amulet that dangles over a Sick Chicken T-shirt. As he walks, wearing clogs, an overloaded key chain swings from his dad-style jeans.

"I'm the only gay man I know," he says, peering through tinted glasses, "who doesn't care about how I look."

Stationed in Thailand during the Vietnam War, Brinsmead wound up with a headful of crazy stories and an I-just-don't-care approach to life. After twenty-seven years, he's about to retire as the physics department's chief technician.

"Actually," he tells me, "I'm not retiring so much as I am being retired. More on that later."

After Phaneuf offered to let Taylor share his lab, he asked Brinsmead if he'd team up with him to mentor Taylor. Phaneuf said Taylor had exceptional talent and was focused and intelligent. But Brinsmead hesitated. He thought about the numerous technical hurdles that they'd need to overcome to build a working fusion reactor, something that hadn't been done at the university before. The odds of success weren't very high. Winterberg was right about one thing: Taylor was biting off more than even the average PhD could chew.

I head to the counter to grab a shot of espresso for myself and some hot chocolate for Taylor. When I return to the table, Taylor and Brinsmead are engaged in an almost impenetrable conversation, throwing out acronyms like SAGE, Z-pinch, and EMFs. I notice that with Brinsmead, Taylor actually stops talking long enough to listen.

"Heard that yellowcake you guys made is hotter than hell," Bill says.

"Yep," Taylor says, "real snotty."

"If it got loose . . ."

Taylor finishes Brinsmead's sentence. "We'd have ourselves an EPA-regulated problem."

We drink up and then jump into Brinsmead's military-surplus electric van, which he's customized. With high-pitched Thai pop music blaring from the sound system (Brinsmead and his Thai partner co-own a house in Bangkok), we drive to the university, passing the Pneumatic Café, which, I will later discover, has a sandwich named after Brinsmead: the Billzilla.

"We've had electric vehicles in my family since 1910," Brinsmead says, "starting with the Baker electric car." Brinsmead built his first EV in 1973 while stationed in Sattahip, Thailand, during the Vietnam War, using parts scrounged from crashed B-52s and bombed-out jeeps and forklifts. "I called it the kamikaze golf cart, and I shipped it home in a big container, along with a lot of other stuff."

Brinsmead grew up in Reno and spent his teenage years "building crazy things." He'd ride to a scrapyard with a basket on his bike, maybe pick up a four-stroke engine or discarded city signs, which he would rehab and sell at school.

"One day I got called down to the principal's office, and the cops were there. I just showed them the receipts and they started laughing."

Brinsmead says he wishes he'd had the kinds of opportunities Taylor has at Davidson. "Grownups would pretty much ignore smart kids who wanted to work with their hands, so we turned into little rebels." One of his teachers did convince him to enter the regional science fair, though. He built a laser out of an old TV and neon signs.

"But I got beat by a girl with a really lame experiment involving guinea pigs and vitamin C. The difference was that hers was well documented."

Brinsmead's father flew B-24s in World War II, and his grandfather ran a bakery car on a troop-supply train in World War I. "But my dad didn't want me going to Vietnam, so he used his connections to have me serve in the coast guard in Lake Tahoe. I was so rebellious, I went down and joined the army and told 'em, 'Send me to Nam.'"

After he scored high on the entrance test, the recruiter told him he could choose any training he wanted. He went to Thai language school and ended up training the Thai army, repairing equipment, and recovering aircraft that had crashed in the jungle.

Once we reach the physics building, our first stop is Brinsmead's office/workshop, which is cluttered with antique physics equipment and other oddities: an ancient cinnamon roll hanging from the ceiling, a bat in a vial, a jarred bottle of dust from the playa at Burning Man.

Brinsmead's irreverent demeanor has often put him at odds with university administrators. "Bill is seen as a bit of a loose cannon," says Phaneuf, "which is something he makes no effort to thwart." But Brinsmead's salvage skills translated well to civilian work at UNR's physics lab. According to Phaneuf, it was Brinsmead's ability to find essential equipment on a tight budget that enabled the university to properly launch its physics department. "To researchers like me, having someone like Bill is essential."

Brinsmead developed a knack for knowing which research and development labs were closing or changing over to new equipment and for showing up at just the right time to claim their surplus high-tech gear. "I always went to the back door and tracked down the facilities manager. He has a problem: too much stuff. You point to your diesel trucks and tell him, 'I'm willing to take everything,' and you've just solved a *big* problem for him."

Some of Brinsmead's skills have rubbed off on Taylor. "The other day," Taylor tells him as we walk down the hallway, "I asked a hospital if they had anything they could donate. I got some detectors with carbon-14" (a rare radioactive isotope).

"The storage room is always in danger," Brinsmead says as we approach. "The professors can't do without it, but the administrators wish they didn't have it, because it puts the equipment out of their control. If you control the equipment, you control the research."

As Brinsmead tries to locate the key, I can see a look of anticipation, almost awe, come over Taylor's face

"I remember coming here for the first time with Bill and Ron," he says. "It was one of the most exciting moments of my life."

Brinsmead finds the key and pushes the door open.

"First time I met Taylor," Brinsmead says, as we walk in, "I have to say it was quite a surprise to hear these Nobel Prize–level words coming out with that Arkansas accent. I couldn't figure out at first if he was just filled up with buzzwords or if he really knew what he was talking about. It didn't take long before it became apparent that he did."

Brinsmead and Phaneuf took Taylor around the storeroom, crowded with a geeky abundance of high-precision gear.

"The first thing I spotted on the shelves was a neutron detector," Taylor says. "Bill handed it to me and he says, 'Have it.' "

After a few minutes of rummaging, Brinsmead found what they'd come for: a salvaged high-vacuum chamber made of thick-walled stainless steel, with two view ports. Fabricated with great precision for the semiconductor industry, it had likely cost tens of thousands of dollars new. Lightly used and liquid-cooled, with precisely machined fittings, the piece would be perfect for the fusor's plasma chamber, which would have to withstand extraordinary levels of heat and vacuum.

Taylor hesitated, then stammered, "Can—can I use it?"

"What do you think, Bill?" Phaneuf said.

"I can't think of a more worthy cause," Brinsmead said.

But Brinsmead was still on the fence about the project. Phaneuf had told Tiffany and Kenneth that he was hoping to get Brinsmead involved. "Ron thought he'd be great for Taylor," Kenneth says. "He said he was well liked in the department and could figure out just about anything. And so we invited him over for dinner."

"I was a little nervous," Brinsmead says. "I guess I was stereotyping. I mean, I hear that accent and I'm thinking Rebel flags and hicks and hate rallies. But they were nothing like that; there was no prejudice or kooky religion. They were smart and open-minded and they had two little geniuses whose talent they needed to cultivate."

Though they made an unusual combination, Brinsmead hit it off with Tiffany and Kenneth, and Joey too. As he was driving back home that night, Brinsmead says, he recalled his own boyhood, thirty years earlier, when he was bored and unchallenged and aching to build something really amazing and difficult (like the laser that he eventually did build) but dissuaded by most of the adults who might have helped him.

"I decided that night," Brinsmead says, "that I'd help Taylor build his star."

21

A FOURTH STATE OF GRAPE

JAN AND BOB DAVIDSON spend a lot of time at the academy they founded. They drive down from their home in Lake Tahoe and often time their arrival to coincide with class changes. Side by side, they move through the throng of students like proud grandparents. Bob plays the gruff jokester; Jan, who has a motherly air, squeezes in a dozen hugs in the few minutes between classes.

"Look at them," Bob says as we maneuver through groups of students laughing and jostling their way toward classrooms. "We cling to this idea that children are better off socially with children who match them in age — despite decades of research that tells us that's not true, that kids are comfortable with other kids at their own intelligence level."

After spending a few days at Davidson Academy, watching the students thrive in an environment with individualized learning plans and the opportunity to pursue their passions as far as they can take them, I can't help feeling that this is the way every school should be. Davidson students may be outliers, at the top of the talent curve, but it's hard to imagine that any child, whatever his or her gifts, wouldn't bloom with this kind of pedagogy.

When I mention this to Jan and Bob, they give me looks that would be accompanied, if they were the same age as their students, with the word *Duh!*

"It's not like we've invented a new paradigm," Bob tells me. "We just looked at what works best in education and we applied it to gifted kids. We didn't listen to the people who've done education wrong for a hundred years." The approach is commonsensical, he says, and backed up by research. It works whether a child has a talent for intellectual learning or for some other field, be it athletics, art, or entertainment.

"It's simple," says Jan. "First, recognize their talents, then try to find them opportunities that support them as individuals. Then, don't put limits on them. Let them bloom, let them pursue what they want to pursue."

When the Davidsons were planning their academy, they approached the nation's top gifted-and-talented-education researchers for advice on best practices, and they built much of what they learned, in addition to some of their own innovations, into their school. Though the academy has more flexibility than many public schools, much of what works at Davidson could be applied to just about any school:

Individualize learning: Meeting each child's unique needs is the keystone of the Davidsons' educational philosophy, and it should probably be part of every school's policy—though it clashes profoundly with the dominant model of one-size-fits-all education. "Basically, we teach deeper," says Jan. "It's not a race to check off courses, though it is very goal focused. These kids get big ideas, and we want to help them make them come true." Each student at the academy has a learning plan that's refined weekly in advisory meetings, every semester with an academic coordinator, and yearly in what's-next meetings with Harsin. That intensity of contact would strain the resources of large public schools—but then, so do the behavioral problems of students who are underchallenged or struggling to keep up.

Enable acceleration: "Don't hold them back," Harsin says. "If a ten-year-old is ready for a calculus class, arrange it for her." Harsin groups core classes by period—all math classes during second period, all English during third, et cetera—so students can easily move up or down according to ability. In large public schools where students fall more widely along the learning continuum, it can be trickier to create options for children who are ready to take on additional challenges. The best schools and districts strive for flexibility, involving families and finding creative ways to extend curricula and connect students with mentorship opportunities, online programs, transportation to advanced classes at other schools or colleges, or other solutions.

Promote dual enrollment: The Davidsons parked their high school on a university campus, making it relatively easy to meet the needs of students like Taylor, who was ready to do PhD-level lab work in nuclear physics. Many high-schoolers are ready to start taking college classes; dual enrollment allows youngsters to take classes in both middle and high school or high school and college.

Accept and celebrate diversity: Although Davidson Academy isn't racially diverse, the student body includes a significant number of children who are what the gifted-education community calls *twice exceptional.* Some have autism-spectrum disorders; others have physical handicaps and emotional challenges. "We had to figure out how to meet needs that weren't purely academic," says Harsin. Just as important, Davidson students (many of whom were labeled by peers as "different" in previous schools) are encouraged to be who they are. "There are kids there who would, honestly, have a hard time in most other environments," says UNR physicist Bruno Bauer, a Davidson parent. "At Davidson they don't have to pretend."

Colleen Harsin faces several unique challenges in educating some of the world's smartest kids. Not least among them are the intricacies of managing 130 personalized learning plans. "Customizing the learning experience was much harder than we imagined," she says. "We'd assess them and get things figured out, but by the time they got here, they were in a different place. They learn and develop so quickly that it's always a moving target."

Harsin also had to find a way to manage parents who had spent most of their children's lives advocating for them and, sometimes, pushing them in ways that weren't healthy in the long run. Students and faculty are quick to notice the influence of helicopter parents. "Yep, we've got quite a few of them here," says Ikya Kandula, one of Taylor's closest friends. "My parents were very driven, and they pushed me into one gifted program or another most of my life. I eventually found my own motivation and my own path—but only after my dad wanted me to be a doctor or a computer engineer and I said no. But a lot of parents here are very pushy."

Harsin says she's learned to work with two kinds of parents at Davidson: those who are pushing their child along, and those who, like Tiffany and Kenneth, are being pulled along by their child.

"There are parents who have their kids up at nine p.m. studying, prodding them," Kenneth says. "We're not like that; we're far less focused on grades."

Harsin also sees many parents putting their child on a pedestal.

"They're very invested. But smart kids pick their noses too. I'm not here to provide global praise because they've rolled out of bed and are still breathing. 'We don't issue pedestals,' I say."

In fact, one of the secrets to raising and educating supersmart kids may be not to tell them that they're smart. Harsin has adopted the growth-mindset techniques developed by Carol Dweck, the Stanford psychology professor. According to Dweck, who authored *Mindset: The New Psychology of Success,* praising children's intelligence or talent can backfire because it puts them in a fixed mindset; a child can become afraid to take on challenging projects since failure could call into question his image of himself (or his parents' image of him) as a whiz kid. These children become increasingly discouraged when encountering difficult problems and are more likely to lie about poor scores. Dweck's research with children has shown that those who are praised for *effort,* rather than talent or outcomes, tend to develop what she calls a growth mindset; this creates a passion for learning and a sense that effort, rather than approval, is the key to creating knowledge and skills. Growth mindset also makes children more likely to persevere when they fail, because they don't believe that failure is a permanent condition.

"You focus less on the achievement itself," Dweck says, "and more on what the child did to make the achievement happen. It's not, 'I'm so proud of you for building this.' It's 'I've noticed that you're really good at getting resources and that you took on a challenge that wasn't easy.' You focus on strengths and what they did to make their achievements happen."

The issue of motivation is one of the biggest debates in gifted education and in education in general. In the past few years, interest has grown in the concept of *grit,* which University of Pennsylvania psychologist Angela Duckworth defines as perseverance and sustained interest in long-term goals. A former New York City schoolteacher who pursued a PhD to study how students learn from a motivational perspective, she discovered that grit is usually unrelated or even inversely related to measures of talent and intelligence. "Talent doesn't make you gritty," Duckworth says. "Our data show very clearly that there are many talented individuals who simply do not follow through on their commitments."

But do people have to be born with grit, or can it be cultivated? Is there a way to teach kids a solid work ethic? Certainly, grit and persistence are critical to long-term success, but they can be difficult to sustain when the task, the goal, or the subject is uninteresting.

"Especially in domains where the acquisition of expertise is by no means fun," notes Dean Simonton, "it's easier to discourage than encourage. It's difficult for extrinsic incentives to produce an intrinsic fascination."

What's missing in most of modern education is a systemized attempt to align schoolwork with students' particular interests. But that's where the magic happens at Davidson Academy.

"The real problem with school back in Arkansas was that I was bored," Taylor tells me one day in Harsin's office. "They wouldn't let me do the things I was interested in. Here, they override the bureaucracy, and if they don't have the resources, they let me go to the university. I can go as far as I want with the things I want to do. Where else could I do that?"

Davidson Academy was founded with the goal of giving students the freedom to explore and the support of teachers who could offer novel, complex, and comprehensible yet challenging material, day after day.

Philosopher and educational reformer John Dewey was among the first to pick up on the essential linkages between interest, curiosity, and effort. In the early twentieth century, Dewey argued that because interest is what drives active learning, "willing attention" is far more effective than "forced effort."

"If we can secure interest in a given set of facts or ideas, we may be perfectly sure that the pupil will direct his energies toward mastering them," Dewey wrote. In contrast, he noted, an education based on forcing children to expend energy unwillingly only results in a "character dull, mechanical, unalert, because the vital juice of the principle of spontaneous interest has been squeezed out."

New studies have reinforced Dewey's theories, showing that learning is optimized when students have an emotional interest in material and experience enjoyable feelings (such as absorption, fascination, and excitement), and when they feel that the activity is related to their lives and identities. Other recent research suggests that emotional interest and enjoyment can actually make learning more efficient and facilitate creativity, especially when combined with environmental factors such as freedom, support, and positive challenges. Having high emotional interest can also trigger a broad, exploratory frame of mind, a state of flow in which the activity feels effortless. Research shows that that sort of enjoyment can even buffer the depleting effects of mental effort and boost one's fortitude to persevere in the face of setbacks.

Though Davidson Academy's encouragement of highly specialized endeavors wouldn't be right for everyone, it might have been a good choice, had it existed, for some of history's most innovative mavericks, iconoclasts, prodigies, and outsiders. Wagner, Picasso, Michelangelo, Newton, and Einstein were single-minded in pursuing their passions and often at odds with the dominant culture, which, then as now, did not reward youngsters who were quirky, creative, or recalcitrant.

As a student, Einstein was known for his curiosity and rebelliousness; one teacher, Hermann Minkowski, said young Albert was a "lazy dog" when it came to studying mathematics. At root, he was just like any other child, driven by the desire to explore and create, an impulse that is very often at odds with the pressures toward conformity and parental, academic, and societal interests. Beethoven, an impulsive and rambunctious child who exhibited little hint of genius during his school years, would probably be drugged in one of today's ordinary classrooms — or he might just sit in the corner with his smartphone, drugging himself.

Of course, Einstein's and Beethoven's kind of raw genius is very rare — one in several million, Lubinski estimates — and it can't be created by any school. "The more exceptional you become the less likely parents and schools can provide environments that accommodate those individuals," says Lubinski. "You best approach it by providing optimal environments to kids to learn at their desired rates, and not topple the passions of the super-achievers." In other words, you want to be attentive up to a point, help students find their passions and cultivate creativity, then give them the resources they need and get out of the way.

Sometimes, experiential learning at Davidson can go in unexpected directions. One night, Taylor read that it was possible to create plasma in a microwave oven, using grapes. "The idea is that if you cut the grape exactly in half and place it precisely in the center of the oven and bombard it with microwave radiation, you can create a cloud of plasma," Taylor says.

And so he decided to try it — in the school cafeteria. Taylor and his partners in crime, two friends with a similar interest in crazy science experiments, set it up. They cut the grape in half equatorially, leaving a bit of skin attached to act as an antenna for the microwave energy.

"A grape is just about the right size for microwaves to interact with and heat that region very intensely," Taylor told the crowd of kids circled around him as he placed the grape halves on a plastic plate and then positioned it in the middle of the oven. "If we get it right, the water and carbon will heat up, and the high-energy microwaves will strip the electrons off, creating the state of matter known as plasma."

As he closed the door and pressed the start button, heads crowded in, straining for a look. Sure enough, within a few seconds a wispy blue-white cloud—the plasma—began to form over the grape halves. "Whoa!" kids gasped. "Look at that!"

"Behold," Taylor announced, "the fourth state of matter!"

Then the experiment took an unexpected turn. There was a loud, zapping noise followed by sparks and a flame. The plastic plate had ignited, and acrid smoke began to fill the stove and then the cafeteria. They put the fire out quickly, but within a few minutes, Taylor and his friends were sitting in Harsin's office. Harsin didn't even have to ask who the ringleader was.

"How was this a smart idea?" she asked, using one of her favorite phrases for the aftermath of student-induced incidents.

"Well, it was a pretty good idea," Taylor said, "right up till the plastic plate. We should have used glass."

"I made a new rule that day," Harsin tells me as we look out her office window at Shenanigan Central. "No more science experiments in the cafeteria."

She laughs, watching Taylor clowning with Sofia, Ikya, Biff, and a few other kids. He backs up and spins into a sort of helicopter dance—Jerry Lewis–meets–Whirling Dervish—without a trace of self-consciousness, making the other kids double over with laughter.

"Can you imagine him in a normal school, bored by his grade level, not meeting other kids like himself?" Harsin asks. "Tuning out in class and thinking, *I could be writing grants and building a nuclear fusion reactor and developing applications for it.*"

Taylor was flourishing, making new friends, getting going on his fusor project. But Joey, increasingly, was struggling; he missed his friends back in Arkansas. For Joey, being around smart people was nice, but it was far less important than being around *his* people. Whereas Taylor could, from the age of eleven, tell you his career plans—applied nuclear physics—Joey had no idea what he wanted to be.

In fact, Taylor's focus and self-motivation was so ingrained that he had trouble imagining that others could possibly lack it. One day, Taylor asked his mom, "What's Joey going to do?"

"He doesn't know," Tiffany said.

"What do you mean?" Taylor said. "How can Joey not know what he wants by now? He has to know."

"Taylor just doesn't get it," Tiffany later tells me, "that everyone doesn't have a passion by age ten and shoot for stars."

One night, Joey asked Tiffany, "Mom, is everyone supposed to have a dream, like Taylor? 'Cause I don't have one."

"I told him that not everyone knows, that I didn't know when I was in high school or even in college," Tiffany says. "And I told him that it's okay not to know."

22

HEAVY METAL APRON

FOR TAYLOR, SUDDENLY, things were *happening.* In contrast to Winterberg's inflexible, theory-based approach, the people Taylor was meeting now actually *did things* and *built things.* Within a few days, Phaneuf had cleared a generous work area for Taylor at one end of his own high-tech physics laboratory. Brinsmead brought Taylor to the physics machine shop and introduced him to the department's engineers, machinists, and technicians, including Wade Cline, a metallurgy and welding whiz who would become a good friend.

Most afternoons after school, Taylor would walk over to the Leifson Physics Building with his chaperone from the university and descend to the subbasement, where he'd spend the next two hours. On Fridays, when he had just one class at Davidson, he'd spend most of the day at the lab. Phaneuf and Brinsmead checked that Taylor was badged with a dosimeter the whole time, and they outfitted him with a big lead apron, like someone would wear for a dental x-ray, that came down almost to his shoes.

A physics lab can be a lonely place, and Phaneuf appreciated Taylor's energetic company—in fact, Phaneuf says, the late afternoons with Taylor quickly became his favorite part of the day. "He's a happy guy to start with, and why wouldn't he be? He was doing what he wanted, and a lot of it. He's a doer, a creative person, and just filled with ideas. It's a joy to watch him learn, and hear him think and create."

Taylor is attracted to anything energetic, and there's nothing more energetic than a hot-plasma environment. Phaneuf, one of the world's leading plasma researchers, helped Taylor deepen his understanding of the enigmatic workings of this fourth state of matter while he conducted his own experiments on the snaking, highly complex ionization machine that takes up most of the space in the lab. It's one of only two in the world, both of which Phaneuf built from scratch.

Phaneuf used his machine to conduct experiments on how various elements behave when ionized. His research has informed several disciplines and has helped other researchers understand deep-space nuclear reactions by simulating astrophysical hot-plasma environments.

Manmade low-temperature plasmas light up fluorescent bulbs and neon signs. When electricity flows through tubes containing certain gases, it ionizes and excites the atoms and creates glowing plasma inside the tube. But high-temperature plasmas are rare on and near Earth. Auroras (the northern and southern lights) occur when charged particles from the sun get caught in the Earth's magnetic fields. Spiraling toward the poles, some of these particles reach the upper atmosphere, where they ionize nitrogen and other elements, forming plasma. As these ions recombine with electrons light is emitted, giving us spectacular light shows.

While natural plasmas are scarce on and near our planet, about 99 percent of the known matter in the universe exists in the plasma state. For instance, stars are big balls of plasma that host nuclear fusion reactions. Taylor knew that he'd need to deepen his understanding of high-temperature plasma to achieve fusion; under Phaneuf's tutelage he soon became obsessed with this temperamental state of matter.

The Farnsworth-Hirsch fusor is an ingenious device. "Think of it as a Polish firing squad," Phaneuf says with a laugh. "You've got ions in a circle shooting inward. If all goes well some will meet head-on in the middle, violently enough for the nuclei to fuse together, releasing high-energy neutrons."

Near the middle of the spherical reaction chamber is a negatively charged spherical grid. When the fusor is fueled with deuterium gas and electrified, the high-voltage current strips electrons off the deuterium atoms, converting them to positively charged atomic nuclei (ions) that fly toward the negatively charged inner cage. When the pressure, voltage, and fuel mix are just right, a glowing core of superheated plasma forms inside the cage. Mimicking the core of the sun, its ultra-high density serves as an attractive target for the converging ions.

If the converging ions miss the cage and one another, they fly out the other side. But as they move outward, they come under the influence of the inward electrostatic force, which directs them back toward the center, where they have the chance to meet again. The majority of impacts don't result in fusion, but some are energetic enough to overcome the electromagnetic forces that push them apart. Then the strong nuclear force takes over and the nuclei fuse, releasing their energy in the form of neutrons.

In a fusor, the fusion reactions are driven by particle velocity more than heat, so little energy is lost through light and other radiation. The lack of magnetic fields also lowers energy losses. But a fusor's major limiting factor is the grid itself, which robs charged particles of their energy when they strike it. This hobbles the fusor's capacity to produce power and wastes the energy invested in ionizing and accelerating the particles. The strikes also cool the plasma down and can damage the fine-mesh grid itself.

A Farnsworth-Hirsch fusor is far easier to describe than it is to build. To get it right, Taylor needed to design and construct a system with such precision that it could maintain vacuum pressure several orders of magnitude lower than the vacuum of outer space. He needed to precisely concentrate up to 100,000 volts of electricity and coax his plasma core up to about 580 million degrees within an environment robust enough to sustain and survive those torturous conditions. And to verify his results, he needed to develop a detection system that could accurately differentiate neutrons from other types of radiation and document their presence convincingly enough to overcome the skepticism of the Neutron Club gatekeepers at the Fusor.net forum. Thinking ahead to the process of data acquisition, Phaneuf helped Taylor set up a LabVIEW software system and got him started on coding the system's data-collection functions.

Phaneuf had helped to get Taylor's nuclear fusion project moving by providing the space, critiquing Taylor's designs, and cutting through bureaucracy. As work progressed, Brinsmead grew more involved, to the point that he was there with Taylor most afternoons. Their friendship grew as they brainstormed design ideas, scrounged parts, and "conned the machinists to help Taylor do some work on it," as Brinsmead put.

"Bill was pretty much perfect for Taylor," Phaneuf says. "He had the right personality for mentoring. He made it fun, but he could also say, 'Hey, Taylor, wait a minute; before you do that, you'd better check this.' That's especially important in this field, because if you get excited or tired and make a mistake, it can be lethal. Taylor is excitable and somewhat impulsive by nature, so we really worked with him to get him to stop and assess a situation before taking action."

Mentorship can be the key to keeping very bright kids headed in the right direction. The vast majority of those who go on to illustrious careers or discoveries say that having a mentor was an important part of their development. And yet, meaningful mentorship is largely absent in the education of not only the gifted and talented, but all children. That's partly because the process of setting up effective one-on-one relationships is hard to institutionalize. Finding the right person for a student, someone who has expertise and time and the willingness to take a personal interest, can take a great deal of energy and effort. As resources for education have been cut and workloads have increased, it has become harder for educators and institutions to be good mentorship partners.

That shouldn't let schools off the hook, advocates for gifted and talented children insist. "All the AP and summer programs in the world are not enough," says researcher David Lubinski. "Kids respond much better to a mentor; those one-on-one relationships are extremely important and extremely effective if you get the match right."

Phaneuf's and Brinsmead's hands-on brand of science was a good match for Taylor. "You can't buy the stuff in my lab, just like you can't buy a fusor," Phaneuf says. "We have to roll up our sleeves and build practically everything from scratch. So what Taylor was trying to do fit right into the style of what we were doing."

"Taylor was really lucky to find Ron and Bill," says Elizabeth Walenta, the science teacher. "They were able to see this tiny fourteen-year-old for who he was and what he could do, and they could help him do it safely."

But it was Taylor's persistence and initiative, not luck, that got him the guidance he needed after he was shot down by Winterberg. "The right mentoring is essential," says Simonton, "but it's not always easy to obtain. A significant part of talent is the willingness to seek out the best mentors, even when rejection is a possibility."

"I wish UNR would make it easier and get more involved in mentorship with students," says Walenta. "A lot of kids want to do that, and it can really work for a student like Taylor, who is motivated and gregarious and affable. But others are not as persistent. I send some kids over to the university, and they get lost and hit a wall and give up."

"Taylor, you're definitely a hands-on guy," Brinsmead told Taylor one day as they inventoried the hundred and fifty or so parts Taylor had brought with him from Arkansas. "Not too many kids are like that anymore; they don't want to get their hands dirty and try things like you do.

"But the problem is," Brinsmead continued, "the stuff you want to build is so incredibly expensive."

According to amateur fusion pioneer Tom Ligon, acquiring the parts to build a working fusor is the second-biggest stumbling block, after acquiring the expertise. "Unfortunately," says Ligon, "most of the people who start on a project like this are too poor to leap in and build something quickly. The people who stick with it are typically above normal in terms of their motivation and tenacity."

As it turned out, very few of the parts Taylor had brought from Arkansas would prove up to the task. But when it came to scrounging, Taylor had one of the best mentors imaginable. Brinsmead downplays his influence. "I may have taught him a few tricks, but let's face it: there isn't a lot that Taylor can't talk someone into — or out of."

Taylor made the rounds of his growing group of friends and supporters in the physics department and beyond, scaring up more Con-Flat fittings, finding a source of finer-gauge tungsten for the fusor's grid, taking an electrical control panel off the hands of a professor who no longer needed it. He and Kenneth and Brinsmead drove out to a closed bank to confirm rumors of a video camera left behind, and Kenneth used his Rotary Club connections to track down someone who could provide permission for them to take the camera. They visited one of Brinsmead's old surplus haunts and found an IBM disk-drive frame from the late 1970s that could support the machine.

Taylor had a head start with his stainless-steel chamber, but locating other critical parts at prices he could afford was a major challenge. It would be crucial, as Taylor had already learned, to create and sustain a high vacuum. He had started with refrigerator compressors wired to run backward, then advanced to single-stage vacuum pumps, then to the oil-diffusion pump from the astronaut's mass spectrometer. But both Brinsmead and Phaneuf doubted that the oil-diffusion pump was up to the job. Taylor, making the rounds of the physics department, heard that a professor had purchased a two-stage turbo-molecular pump for an experiment but a grad student had let it roll around in the back of a truck and some of the parts had fallen off, and now it was inoperable.

"That was a twelve-thousand-dollar bearing-stabilized turbo pump,"

Phaneuf marvels, "and somehow, Taylor ended up with it because someone didn't want to bother to fix it." Taylor was able to buy or build all the missing parts for the turbo pump, which could spin at jet-engine speeds. Once he got it working, he teamed it with the diffusion pump to create a two-stage system strong enough to reduce the pressure to a workable level.

Even though Taylor eventually found his way around Winterberg, the encounter had been influential, forming the roots of Taylor's growing distaste for theoretical physics. It would cement his devotion to hands-on invention as he became less interested in the paths of the Einsteins and Heisenbergs and more focused on putting their discoveries to work in the manner of Tesla and Farnsworth. "I want to apply the physics theory that's been discovered," he'd say, "and use it to create something new to solve real-world problems."

As a result of the clash with Winterberg—who had told Taylor to master calculus and then work his way up through the theoretical physics courses before attempting anything hands-on—Taylor "dragged in a fair bit of defiance" to his math classes, according to teacher Darren Ripley. Taylor doesn't disagree. "I had a chip on my shoulder," he says, "and some of it is still there." In New Mexico, Taylor told Willis that he didn't believe calculus was "all that useful." When Willis raised a skeptical eyebrow, Taylor responded unyieldingly: "Carl, I don't want to be a theoretician!"

And yet, without a thorough understanding of physics theory, it would be impossible for Taylor to build a fusion reactor that could mimic the power-generation process of the sun, impossible to understand the behavior of the subatomic particles that would, he hoped, collide and fuse in his reactor, releasing their neutrons. Winterberg was right: to be a physicist, whether theoretical or applied, Taylor would need to stand on a solid scaffolding of high-level mathematics.

Ripley's job—to help him build that scaffolding—was made tougher by the fact that Taylor, like most of his incoming students, had been "doing math in a way in which they don't know where the concepts come from." It's a criticism I would hear from a lot of experts in STEM education: Mathematics continues to be taught as if it has no history and very little future application to what kids see as relevant to their own interests and career plans. "It's usually just drill and kill," says Ripley, "and that's what kills interest in math."

But Ripley is a talented teacher, and his small class sizes allowed him to know his students well enough to make the material applicable to each student's particular passion. Ripley also understood Taylor's dominant learning style—he makes sense of things by talking about them—and so he would often turn the discussion over to Taylor to connect a mathematical concept to some phenomenon of radiation.

"Tay, why don't you tell us why this is useful?" Ripley would say. "Then," Ripley tells me, "once he got going, it could get pretty entertaining, and you could see the light go on. Suddenly, it's relevant to him. You could see new pathways forming, wheels turning, deeper thought processes."

Taylor and Phaneuf both agree that Taylor got a lot out of Phaneuf's upper-division nuclear physics class—"although," says Phaneuf, "some of the math was beyond him." Phaneuf suggested that Taylor take UNR's nuclear chemistry class with Dennis Moltz, a guest professor from Berkeley.

According to Moltz, whenever he asked the class a question, Taylor would blurt out the answer. At Davidson—where Taylor's peers shared his intellectual intensity—that wasn't a big deal. But in the university environment, it didn't go over so well. Moltz had to take Taylor aside and ask him to restrain himself.

Taylor was exhibiting another common trait of highly gifted children: intellectual overexcitability, or intellectual OE. According to psychologists, those high in intellectual OE are intensely curious and usually able to concentrate, engage in prolonged intellectual effort, and tenaciously problem-solve when they choose. But their extreme eagerness can cause social difficulties when they become so excited that they interrupt at inappropriate times or grow impatient with others who can't keep up with their intellectual pace. Polish psychologist and physician Kazimierz Dąbrowski, whose studies of gifted schoolchildren led to the theory of intellectual OE, says that people with high intellectual OE tend to have powerfully creative perceptions of the world.

But not all creativity is created equal. Mihály Csíkszentmihályi and other psychologists and creativity researchers distinguish between what they call big-C and little-c creativity. Little-c creativity includes garden-variety problem solving and the ability to adapt to change. Big-C creativity, which is far more rare, occurs when a person solves a problem or creates something original that has a major impact on other people.

"At the little-c level, creativity implies basic functionality," says Simonton. "And at the big-C level, it's something that we give Pulitzer and Nobel Prizes for."

A former president of Caltech contended that one truly excellent scientist is more valuable than a thousand very good scientists. "Indeed, it seems unlikely that 1,000 average scientists could have produced the General Theory of Relativity," wrote aerospace pioneer Norman Augustine, "no matter how much time they were allocated. Nor could 1,000 average writers have produced Shakespeare's works. Nor could 1,000 composers have created Beethoven's music."

One challenge for gifted kids, says Ellen Winner, is making the leap from being a gifted child to being an influential creator. "The skill of being a child prodigy is not the same skill as being a real big-C creator. A child may be extremely good at mastering something, but most kids doing calculus at eight aren't necessarily creators. That's a very different kind of mind. Most prodigies don't become creative geniuses, they become experts. Only some manage to move beyond that to a more reflective sense of direction."

23

BIRTH OF A STAR

SEVERAL MONTHS INTO his fusion project, Taylor was still missing a few parts, but he had, thus far, spent less than two thousand dollars on his fusor, even though it contained tens of thousands of dollars' worth of precision parts. Some had been custom made in the physics department's state-of-the-art machine shop, others scrounged from equipment rooms, surplus suppliers, UNR professors, and friends and colleagues around the country.

But some of the more specialized (and critical) parts were proving harder to find. The electrical insulator Taylor had finagled didn't have enough insulation to keep it from ionizing the air around it (much like high-voltage power lines do, a phenomenon that's often noticeable in wet weather). Whenever he cranked the voltage up, the power supply would start arcing and sputtering.

And so Taylor took his quest farther afield, again working his advantages of youth and novelty. Taylor charmed Willis's boss, whom Taylor had met on one of his visits to Albuquerque, into letting him have an oil-bath insulator worth several thousand dollars for a hundred dollars. The next time Taylor and Kenneth went east, they stopped in Albuquerque on the way back and picked up the green, oil-filled drum, which immediately fixed the problem.

One of the most vexing challenges for fusor builders is the grid, the spherical wire shell that cradles the plasma. A fusor's grid must be as transparent as possible, since every particle that hits it will lose energy. The grid needs to be able to withstand the intense heat around the plasma core, which will, if all goes well, probably be the hottest microenvironment on Earth during the minutes when the fusor is running. Since the grid must be robust enough to stand up to the hits by subatomic particles, most achieve 90 to 95 percent transparency—which means that on every pass, up to 10 percent of the deuterons will collide with a grid wire and lose their energy.

Taylor wanted to do better; he came up with an idea for a grid design that would be about 98 percent transparent and took it to Wade Cline in the physics department's machine shop. A former Lawrence Livermore engineer who had been on the Greenwater project (a top-secret Star Wars concept for an x-ray laser powered by a nuclear blast; it was abandoned in 1992 after the military had invested a billion dollars in it), Cline was known as a miracle worker when it came to machining extremely high-tolerance doohickeys for departmental experiments.

Cline and Taylor sat down and brainstormed an approach that would come close to Taylor's goal. But to make it work, Taylor needed to find some high-purity tungsten, which has a high melting point and is easier to work with than tungsten with even minor impurities. Taylor's sleuthing eventually led him to Bauer, who had helped establish UNR's Nevada Terawatt Facility, a laboratory on the outskirts of Reno set up to study the behavior of matter subjected to conditions of extreme temperature and density. At the Terawatt Facility, Taylor found the super-pure tungsten wire he needed. Then he and Cline used a CAD program and CNC mill to design and machine a hub that could secure intricately shaped pieces of wire and mate them with the feedthrough stalk.

The problem with tungsten, though, is that it gets brittle when superheated. The first few times Taylor and Cline subjected it to high-heat conditions in the fusor, sections of the wires began to glow, and then the ultrathin grid collapsed under its own weight. Taylor and Cline kept tweaking the design and remachining it, going through several iterations until they finally hit what seemed like the sweet spot between strength and transparency.

By the time Taylor arrived in Reno, he had built up a mostly self-taught base of knowledge in at least twenty fundamental fields of science and engineering, including nuclear physics, chemistry, radiation metrology, and electrical engineering. Now he was learning about mechanics and metallurgy, tool handling and hands-on experimental methods—things that he'd be hard-pressed to learn at any school. He spent a lot of time in the machine shop, learning by watching the technicians work on other projects. With the help of his growing circle of mentors, Taylor began adding expertise in plasma physics and metrology, electronics design, and vacuum technology.

"I was learning so much," Taylor says. "The biggest thing was technique. I came in thinking I knew things, but compared to the guys at the lab I knew nothing about the design and method and construction and operation of scientific equipment. They had all the tricks, and they were willing and able to help me figure out a way through just about any technical problem."

Back when Taylor was working in his grandmother's garage, he says, "it was all about function." He'd use aluminum cans, rubber, the innards of old vacuum cleaners. It wasn't pretty but the devices worked —sometimes. What he began to understand, watching Cline and the other technicians build extremely high-precision machines and parts from scratch, was the inseparability of elegant design and functionality. "I saw that if you really think it through and build things very precisely out of just the right material, it comes out as an incredibly functional and incredibly beautiful piece of art," Taylor says. "It not only works better, it's also beautiful. From then on I was committed: all my reactors and everything I made would not just serve a function; they'd also be beautiful machines."

The UNR physics department gave Taylor more than just access to the infrastructure and expertise he needed; it also gave him the chance to develop camaraderie with much older people who spoke his language. Taylor loved hanging out at the lab, the machine shop, and Brinsmead's office. At fourteen, he was playing at the edges of the major leagues, having conversations that would have been unimaginable only three months earlier. He might walk into the lunchroom and find himself in the middle of a chat between physicists and/or technicians about yet-to-be-discovered heavy elements or whether a certain experimental alloy might work for the ITER reactor's walls or if the neutron howitzer was due for a renaissance.

Taylor could start a conversation about just about anything, and his chain-reacting ideas came out of his mouth almost as soon as they came into his head. One day, he walked into the lunchroom, sat down at a table, and asked, "What if you took radioactive fuel, encapsulated it in buckyballs, and accelerated radioactive decay?" initiating an hour-long speculative debate. Another day he asked, apparently apropos of nothing, "Have you heard about the Museum of Stillborn Babies in Kazakhstan?" —a question that started a long discussion comparing the Soviet Semipalatinsk Test Site in Kazakhstan to its American counterpart, the Nevada Test Site.

For the professors and technicians, having Taylor around injected both amusement and energy into the once-quiet department. "Having a young guy in here with Taylor's enthusiasm was pretty exciting; it was something different going on," says Phaneuf. "The doctoral fellows in particular were always fascinated by him; they became his biggest cheering section. He'd bring them out of their shells. I remember I had one German postdoc who was very curious, but shy. At first, he'd peek around the corner at what Taylor and Bill were doing, but pretty soon his shyness was disappearing and he'd just walk in and become part of the group."

Over the next few months, as work on the fusor progressed, Phaneuf watched Taylor's side of the lab transform from an empty space into a sprawling goat's nest of parts, tools, and other clutter. "Myself, I like things neat," Phaneuf says, "but neither Bill or Taylor are particularly neat, and when I looked over there I tried to keep in mind that a creative mess is better than idle tidiness." Phaneuf typically spoke up only when safety was an issue, such as when Taylor left something in the middle of the floor.

Sometimes, Phaneuf would look over and see Taylor working on his machine in an intense state of concentration, totally absorbed in what he was doing. He'd fallen into what Phaneuf and Brinsmead would call the zone. Often they'd have to call his name two or three times to snap him out of it.

"Time just disappeared," Taylor says. "I'd be so into working on something that I'd think maybe ten minutes went by, then I'd look at a clock and it was three hours later."

"VIP in the lab!" Brinsmead shouted toward Taylor, whose head was wedged between the shelves of the IBM rack as he snaked a patch cord through a thick mass of wires. Taylor disentangled himself and straightened up.

"Oh, hi, Sofia!" he said.

Though he was still lacking a few crucial components, Taylor had begun to test-assemble the reactor, which was now bolted atop the chest-high rack.

"Wow," Sofia said, stopping a few feet away to take it all in. "It's not what I expected."

"What did you think it would look like?"

"I'm not sure. I thought it might be . . ."

Brinsmead, sitting on a chair at a nearby workbench, couldn't resist. "Bigger?"

"It's definitely complicated-looking," she said, moving closer and peering into the glass portholes.

The one area that wasn't moving forward in Taylor's life was romance. Taylor had made no secret of his attraction to Sofia, but things had gone nowhere—even after she broke up with her boyfriend. And yet, there was no hint of awkwardness when they saw each other. They hung out regularly, often with Ikya and other friends. And as far as Taylor or anyone else could tell, that was as far as it was ever going to get.

"So, look inside the window there," Taylor told her. "Can you see that circle of tungsten fingers? That's the grid, that's what's going to cradle to plasma field. What I'm going to do is basically mimic the conditions inside the core of the sun and try to get the atoms to fuse so they'll release their neutrons."

"Wait, back up," she said. "You're going to mimic the conditions inside the core of the *sun?*"

"Yeah," Taylor said. "But the sun's got two things that we haven't got here on Earth, unless we build them. The first is, it's got really high temperatures at its core. The second is, it's got massive gravity"—here Taylor spread his arms wide and then drew them in—"which pulls the hydrogen atoms together at super-intense pressure."

Sofia peered through the view window.

"Go ahead, you can touch it," Taylor said. "It's not plugged in."

She walked around the fusor, tugging on hoses, touching the stainless steel.

"How hot will it get in there?" she asked.

"The plasma in the grid gets about forty times hotter than the sun," he said.

She stepped back. "Whoa. Seriously?"

Brinsmead, watching and listening while he pretended to attend to something else, smiled.

"But," Taylor added, "[most of] the particles won't get that hot; they'll only get up to about a million degrees Fahrenheit. That's the beauty of a fusor. If we get a real good vacuum and really high voltages, we can fire those deuterium ions at super-high speeds into the middle of the plasma. Then they collide and fuse together, and some of their mass is released as energy. You know, E equals MC squared and all that stuff."

"Wow, Taylor," Sofia said. "It's like you're Einstein or something."

"Sofia!" he said, exaggerating a disappointed look. "I'm an *applied* nuclear physicist. I take basic science, like Einstein's theories, and add engineering to it. Then I make something useful."

"As useful," Brinsmead chimed in, "as a star in a jar."

"Well," Taylor said, "I *will* be one of the few people on earth that can birth a star. And I've got all kinds of plans to make stuff with it."

"Is it dangerous?"

Taylor explained that the fusor would produce x-rays and neutrons, "which are lethal if you're standing in the wrong place for very long." He showed her where they would soon set up a wall of lead blocks to shield the control-panel area, and he showed her the camera they'd use so they'd be able to see what was going on inside the reactor.

"She seemed pretty impressed," Brinsmead said after Sofia had to run to catch her ride home. "I don't know, Taylor. Maybe you've still got a chance with her."

"Nope, I kinda doubt it," Taylor said. "I think I'm just going to have to accept that we're just going to be really good friends. Which is okay by me."

A few days later, another VIP visited the lab. When Elizabeth Walenta stopped in, Taylor had circled the machine with yellow radiation-warning tape and signs. He and Brinsmead were building the wall of lead blocks, and the control panel was taking shape.

"Wow," Walenta said, "looks like you're getting close."

"Still need a couple more pieces," Taylor told her, "then we can fire it up and start making our test runs."

"You know," she said, addressing both Taylor and Brinsmead, "next week I've got someone coming in, George Ochs, to do a presentation about the Western Nevada Regional Science Fair. He directs it, and he's helped quite a few kids out."

Walenta circled the machine. "This looks really good," she said. "You may want to think about entering it."

George Ochs was at first completely baffled by Taylor. A retired honors-level science teacher and wrestling coach, Ochs has a husky physique, a round face with thin wisps of hair on either side, a brushy mustache, and friendly blue eyes.

"So, this incredibly skinny kid comes in and he's superexcited," Ochs says. "He's talking about radioactivity and how he's got one of the largest collections of radioactive materials in his garage, and now he's going to fuse atoms and produce neutrons. And I'm thinking, If this is all true he'd be glowing in the dark. So we'd better get this checked out."

Ochs asked Taylor a couple of specific questions about radiation.

Taylor answered quickly, and in detail.

Then Ochs asked him about monitoring and permits.

"He answered confidently and it seemed like he knew what he was talking about. But I thought, wow, how could such a young guy be so articulate about such a highly technical subject? I mean, how many people even know what neutrons are? So I asked him more, and it became apparent that he's got what seemed like basically a PhD-level knowledge base about this stuff."

Ochs had twenty-five years of science teaching under his belt, but he'd never met a kid like this. To verify Taylor's story, he called Walenta, then Phaneuf, then Kenneth.

"It all checked out," he says. "He had all the bases covered. And he wanted to enter the science fair, which was great. After that, I started calling him Neutron Boy."

On a Friday afternoon in late February, Taylor and Brinsmead tightened the last bolts. They were ready to bring the machine up to full power. Moving behind the wall of lead bricks, Taylor flipped the main switch. But nothing happened. It took the rest of the day to diagnose the electrical problem and get everything rewired. As soon as they finished fixing it, Kenneth arrived to pick up Taylor.

Taylor begged his dad to let him stay and make their first full-power run. But the family had plans that night. Taylor would have to wait until next week. It was a relatively minor setback, but to Taylor, the weekend felt like an eternity.

On Monday afternoon, though, he and Brinsmead were back at the lab. They checked the machine over, then powered it up. Taylor increased the voltage to ten thousand, then twenty thousand volts. He looked at Brinsmead.

"Go for it," Brinsmead said.

Taylor twisted the knob a little more. Then, suddenly, there was a small pop, and the instruments went dead.

"I think," said Brinsmead, "we just fried the power-supply unit."

Taylor and Brinsmead repaired the power supply and fired up the machine again. At low voltages it worked fine, but each time they took the power up a notch, another problem would arise. Taylor troubleshot the electrical issues, as well as the pesky vacuum leaks. He replaced malfunctioning gauges, methodically working through each problem.

They pushed the voltage up again, with disappointing results. It was becoming clear that the high-voltage feedthrough Taylor had scrounged was not up to the task of handling the voltages that fusion would require.

A more robust power supply would cost at least six hundred dollars. But doors continued to open for Taylor. The next time Phaneuf was at his Lawrence Berkeley Lab, he asked around and found a colleague who had a power supply he wasn't using that would be perfect for the fusor. "That was probably one of the most critical things," Taylor says. Finally, he had the kind of power he needed to hurl atoms together with enough oomph to fuse them.

In February, once Taylor had the new part installed, he and Brinsmead started making test runs with increasingly higher voltages. After a few successes, they decided to take their first shot at generating plasma. Then they would make the final big push toward fusion. "Next week," Taylor told his family and his friends at Davidson, "could be the big week."

Taylor wanted to fuel the fusor with deuterium right away. "I was thinking, *Why waste time, let's just go for the neutrons.*" But Phaneuf and Brinsmead suggested that they first use argon as a test gas. An inert noble gas that transforms into a bright blue plasma, argon could stand in for deuterium and give them a very visible plasma while they made sure all the systems were running properly.

Taylor worked the instruments, slowly increasing the power and vacuum, then adding in a little argon gas. Brinsmead watched the video monitor for signs of the eerie glow called a Paschen arc, which would indicate the presence of a plasma field. As Taylor brought the power up to ten thousand volts, Brinsmead saw a distinct bright spot appear in the center of the fusor's inner grid.

"I think—yep, there it is!" Brinsmead yelled. But before Taylor could look, it was gone. In a few seconds it appeared again—then disappeared. The plasma field kept blinking out. What was going on?

It took hours, but they finally figured out what was happening: the electricity feeding the grid was short-circuiting between the feedthrough stalk and the case.

A few days later, having fixed that, they powered it up again. Taylor and Brinsmead played around with the system, trying to optimize gas, vacuum, and voltage levels. They cranked it up to ten thousand volts, then fifteen thousand. When they brought the voltage beyond twenty thousand volts and pumped out the gas, the instruments detected x-rays—lots of x-rays—coming through the top view port.

"I'm glad we're behind these bricks," Brinsmead said. He and Taylor watched the video monitor, and in a few moments they saw what they were looking for: a glowing ball of blue-white plasma developed in the middle of the grid. This time, the plasma field didn't go away; Taylor had succeeded in producing a stable and sustained high-temperature plasma.

That night—actually, it was 3:40 the next morning—Taylor sent documentation of the event, including photos, to the Fusor.net forum. The posting, dated March 5, 2009, reads:

> Hey Guys,
>
> Well, its [*sic*] time for some plasma pictures. We played around a little with the system and the gas handling with some argon. It took some fidgeiting [*sic*] with the gas to get a discharge, but not flood the current. We have a few issues, have to hook up the Baratron to see what we're dealing with inside the chamber, and add a small amount of tubing, etc. to the Deuterium cylinder, but I think we are ready to go for fusion in a couple of days. Here are some pictures of the discharge. We were running at about 5–15 KV without very many x-rays, but they started to shine through the top viewport as the gas was pumped out and the voltage went over 20 KV. The pressure was pumped down to I think somewhere around 5e-7 before we introduced gas.

By sunrise, "Welcome to the Plasma Club!" congratulations were streaming in. Willis, the first to respond, wrote: "This will be a neutron generator when it gets a breath of deuterium."

Richard Hull wrote, "Nice work . . . Got a real name so I can put you in the Plasma Club?"

"Yes," the fourteen-year-old replied. "It's Taylor Wilson."

But when they ran the test for Phaneuf, who knew plasma better than just about anyone on earth, the professor's eyebrows came together.

"The shape of that plasma field isn't quite right," he said. "See how it's kind of a funny oblong shape? It needs to be perfectly round. If you can get more symmetry out of your grid structure, that will locate the collision zone closer to the grid's geometric center. You could probably eke out some detectable fusion from it like it is, but you might as well do it right. I think you should redesign the cage and make it more spherical."

Unfortunately, the more spherical version of the grid that Taylor and Cline had earlier machined and tested kept collapsing whenever they brought it up to high voltages. They'd changed the design again and again, with the same result.

Then Phaneuf remembered that he'd used a special tungsten-alloy wire for an experiment he'd done at Lawrence Berkeley. The mix of 98 percent tungsten and 2 percent tantalum was specially formulated to prevent that embrittlement. As luck would have it, the next day Phaneuf was flying down to California. He brought back the alloy wire, which Cline and Taylor fabricated into an almost perfectly spherical grid.

"That did the trick," Taylor says. "It was the key to getting the plasma field right." With the new grid, they could push the electrical discharge up almost to fusion levels and watch as a clean and almost perfectly round ball of blue-white plasma developed in the middle of the grid.

"I don't think I'd enjoy hanging out with Einstein," Taylor said as he led Phaneuf and Brinsmead into the physics building's elevator. "Too theoretical. I'd be more into meeting Tesla, or Farnsworth—who, by the way, invented the television when he was fourteen, same as me."

Taylor had wanted to make fusion before his fifteenth birthday, which was coming up in May. He had been pushing hard for months. By early March he was getting more and more anxious, but there was no question he was closer now. All the parts had been either scrounged or built, then tested and assembled and tested again. Over the past few days he'd made some vacuum-tightening tweaks, put the pumps through their paces, and made several test runs with argon. Now they were ready to move into uncharted territory and try for fusion.

Taylor punched in the secret code to enable the elevator to access the physics department's subbasement. "According to the rules," Phaneuf said, winking, "you shouldn't actually have that code."

Phaneuf unlocked the door to the lab and the three of them entered. Taylor immediately spotted a new canister of oxygen-18, a heavy isotope, next to Phaneuf's desk.

"No, you can't have it." Phaneuf laughed.

Phaneuf had finally broken down a week earlier and suggested a little straightening up. Taylor cleared the area around the machine, moving things to nearby shelves and tables, which further set Taylor's fusor apart from the rest of the lab. The reactor, surrounded by yellow radiation-warning signs, dominated the far corner of the room.

The three of them went over and took a look at it, their eyes adjusting to the light in the windowless room. The reactor was a far cry from the jumble of ill-fitting and ill-behaving parts that Taylor had tried to piece together in Arkansas in his grandmother's garage and his father's Coke plant. It looked elegant, defiant even, a gleaming stainless-steel-and-glass chamber atop a cylindrical trunk bolted to the stout rack and connected to an array of sensors and feeder tubes. Standing on his tippy-toes, Taylor was just tall enough to peer through the window into the reaction chamber, where he could see the golf-ball-size grid of tungsten/tantalum fingers that would, if all went well, cradle the plasma in which the nuclear fusion reactions would occur.

Mounted to the shielding they'd set up around the machine was the video camera they'd scrounged from the closed bank branch, bracketed to the rack and aimed at the reaction chamber. At higher voltages the machine would produce so many x-rays that looking inside the window while it was running—or even standing next to it for a minute or two—would induce acute radiation sickness. Any longer than that would be lethal.

Taylor took Phaneuf through the instrumentation that he'd set up, including a calibrated ion chamber, x-ray detectors, and other survey meters scattered around the room. He had also hooked up a boron-10 trifluoride (BF_3) neutron detector, commonly known in the physics community as "Snoopy." Snoopy would give them real-time information about neutron generation, but because it was sensitive to electromagnetic interference and other forms of radiation, it wasn't sufficient to provide actual proof of fusion. For that, they'd set up neutron bubble detectors, cylinders filled with liquid in which bubbles would appear to decisively confirm the presence of the neutrons and prove that a fusion reaction had taken place.

Taylor removed the argon gas cylinder and replaced it with a cylinder of deuterium gas, using the new valve adapter that Cline had machined. He locked the cylinder into place, then moved to the control panel, shielded behind a wall of lead and polyethylene blocks.

"Eh-hem," Phaneuf said, glancing toward the lead apron and dosimeter hanging on the wall. Taylor smiled and duly put them on, then moved behind the wall, saying, "It's not like we've measured x-rays getting through the lead back here or anything."

"Our number-one job," Brinsmead said, "is to keep you safe."

"Even if it *is* total overkill," Taylor mumbled. Though Taylor had grown taller in the past few months, the heavy apron still extended below his knees, giving him a comical look. "And even if I'm the only one in the room with this ridiculous getup."

"Hey, us old guys are done reproducing," Phaneuf said.

"Or we never got started," Brinsmead added, "and don't plan to!"

By that point, Phaneuf was convinced that Taylor knew everything he needed to know about radiation safety. The apron and the massive amounts of lead shielding were indeed overkill. But Taylor was still growing; his dividing cells were more vulnerable to radiation than the cells of full-grown men. And the test runs had shown that the machine produced a surprisingly high level of x-rays. For Phaneuf, that reinforced the wisdom of continuing the safety overkill.

Apart from giving the apron reminder, though, both Phaneuf and Brinsmead stood back and let Taylor go through all his pre-energizing checks. Then the boy returned to the control panel.

"Okay now, y'all stand back," Taylor said. As Brinsmead and Phaneuf retreated behind the lead-block wall, Taylor shook the hair out of his eyes, reached toward the controls, and flipped the main switch.

24

THE NEUTRON CLUB

TAYLOR TOGGLES ANOTHER SWITCH, energizing the vacuum pumps. He and Brinsmead and Phaneuf watch the gauge as the air is sucked out of the reaction chamber and the pressure decreases to the equivalent of 100,000 feet above sea level . . . then the space station's orbit . . . then the surface of the moon.

Standing just behind Taylor, Phaneuf peers over the boy's left shoulder, Brinsmead over his right.

"Looks like switching out that valve really did the trick," Phaneuf says. "You're getting better vacuum than I've ever seen."

Taylor turns a knob to start bringing the voltage up. The temperature moves past 100 degrees C; at this point, any condensation in the chamber would have vaporized and been sucked out.

Taylor brings up the power on the vacuum pumps even more. "And now," he says, "we're entering interstellar space . . ."

Brinsmead says, "When you were in your grandma's garage and you got the idea to try to build this thing, did you ever imagine you'd be at this point, doing what you're about to do?"

"To be honest, Bill," Taylor says, "I did. I just didn't think it would take this long."

Taylor scans the gauges. "I'm going to open the line and bring in a little deuterium," he says, "and give the grid a little negative voltage."

Taylor knew that achieving fusion depended on getting a just-right balance of vacuum, gas supply, and voltage. While Phaneuf was away, he and Brinsmead made several test runs, experimenting with different combinations. "I think he's really got the parameters optimized now," Brinsmead tells Phaneuf.

"Ten thousand volts," Taylor calls out, glancing at the meter. He cranks the voltage knob a little more and checks the instruments. "I'm going to pour on more fuel to balance the pump.

"Twenty thousand volts now, and . . ." — he glances at the video monitor — "we've got plasma!"

Sure enough, a pale blue cloud of plasma has appeared, rising and hovering, ghostlike, in the center of the grid. Taylor looks at Phaneuf, then Brinsmead, who nods. "Let's go for it," Brinsmead says.

Taylor turns the knob, taking the voltage higher. "I've got it up to twenty-five thousand volts now," Taylor says. "I'm going to outgas it a little and push the voltage up a little bit more." The power supply crackles.

From behind, Taylor looks like a small Oz kind of figure, his hands darting back and forth, checking gauges, pulling levers, finessing dials. He adjusts the pressure and voltage again while Brinsmead and Phaneuf keep their eyes on the video monitor. They can see the tungsten wires beginning to glow, then brightening to a vivid orange. "When the wires disappear," Phaneuf says, "that's when you know you have a lethal radiation field."

The two men watch the monitor while Taylor concentrates on the controls and gauges, especially Snoopy, the portable neutron detector that they've set up just to the side of the lead-block wall, a few feet away.

"We should be getting pretty close to star territory now," Brinsmead says.

Phaneuf squints at the monitor. Rays of plasma dart between gaps in the now-invisible grid as deuterium atoms, accelerated by the tremendous voltages, begin to collide. The blue-white plasma starts to throw off purple sparklets.

Brinsmead, who's been watching the neutron detector, suddenly shouts: "We're getting neutrons!"

Inside the reaction chamber, separated from the outside world by two inches of stainless steel, deuterium atoms are stripped of their electrons and accelerated toward the dense, superheated plasma core at the center. Each second, tens of thousands of these ions collide violently enough to fuse and release tiny amounts of mass-energy as highly energetic neutrons.

In other words, nuclear fusion reactions are taking place.

Taylor smiles ever so slightly and keeps his hands on the controls.

"Let's see what we can do now," he says, cranking the voltage up.

"Whoa, look at Snoopy now!" Phaneuf says, grinning. The detector registers two hundred thousand neutrons per second, then three hundred thousand—"and still climbing."

"It's really jamming!" Brinsmead shouts, watching Snoopy as Taylor nudges the power past thirty thousand volts.

"You're getting eight hundred thousand neutrons per second," Phaneuf says. "Nine hundred thousand now . . . a million!"

Brinsmead lets out a whoop as the neutron gauge tops out.

"Snoopy's pegged!" he yells, doing a little dance. "Someone needs to turn up the range."

Taylor makes a move for it, but Phaneuf grabs him lightly by the sleeve, stopping him. "Taylor, don't go over there, no." The stripped-away electrons hitting the chamber's wall are emerging as x-rays. Phaneuf reaches out with the x-ray detector, takes a reading, and decides it's worth the risk. He darts to the neutron detector and quickly dials up the range.

"Get away from that, Ron!" Brinsmead says, laughing as Phaneuf jumps back behind the protective wall of lead.

Taylor glances at the monitor, where the star in the center of the machine is now glowing so brightly that the surrounding grid has disappeared completely.

"Just a little more," Taylor says under his breath as he nudges the voltage up, bringing the temperature of his reactor's plasma core to an almost incomprehensible *580 million degrees*—some forty times as hot as the core of the sun.

On the video screen, purple sparks fly away from the plasma cloud, illuminating the wonder in the faces of Phaneuf and Brinsmead, who stand in a half-orbit around Taylor. In the glow of the boy's creation, the men suddenly look years younger.

Taylor keeps his thin fingers on the dial. As the atoms inside the fusor collide and fuse and throw off their energy—1.1 million neutrons per second, then 1.2 million—Taylor's two mentors take a step back, shaking their heads and wearing ear-to-ear grins.

"There it is," Taylor says, his eyes locked on the machine. "The birth of a star."

Taylor powers down the fusor, then he and Phaneuf and Brinsmead emerge from behind the wall of lead. "Let's see what Ricochet Rabbit has to say," Brinsmead says, referring to the bubble detectors they've arranged around the fusor.

"A bubble detector cannot be fooled," Phaneuf says as they walk toward the machine. "If you've got bubbles, it means you produced neutrons."

They had bubbles.

Until that moment, Taylor had kept a lid on his emotions, at first concentrating on operating the reactor, and then concentrating on not declaring success prematurely. Now it was clear: Taylor Wilson had become the thirty-second person on the planet—and the youngest—to achieve nuclear fusion.

Taylor finally let himself go.

"We screamed and we shouted and we high-fived each other," Taylor recalls.

"Taylor," Phaneuf remembers, "was dancing on air."

"We all were!" says Brinsmead. "To see the look on his face, after all that effort and trial and error and all we'd been through. Oh, man, that moment really made it worth it for me. And for Ron too—although I think we'd both enjoyed just about every minute of working with Taylor for the past several months. But now it was all really paying off. He'd done it!"

First, Taylor called his parents with the news. And then he called Willis, who was ecstatic.

The neutron output would increase as Taylor tweaked and optimized the machine over the next weeks, but the bubbles in the detector were proof positive that a nuclear reaction had occurred. They would restart the fusor and make several runs that day to accumulate a comprehensive data set documenting the achievement. Taylor took the data and the photos and videos home that night and sent the evidence to Fusor.net.

"I'd seen the kinds of questions that had been asked and the kinds of things that tripped people up in terms of neutron-detection methods," says Taylor. "So we made sure we had rock-solid proof." Taylor e-mailed video and still images and a full technical-data disclosure detailing the fusor's setup, conditions, neutron-detection systems, and witnesses.

The Fusor.net community agreed that his proof was irrefutable: a nuclear fusion reaction had occurred. Once it was confirmed and verified, Richard Hull posted a message on the forum that the Neutron Club had a new member—the youngest fusioneer ever—and congratulations began pouring in from all over the world.

"I was pretty excited," Taylor remembers, "because it was a validation of what I'd set out to do. But I'd always been confident that I'd do it, so in a way it felt less like winning a gold medal and more like buying the [boxing] trunks to start training.

"It was a beginning. I had this tool now, and the world opened up. I could start playing with my neutrons and do all these experiments and find out all kinds of things. I had a whole world of questions, and now I had subatomic particles that I could use to start opening up the answers."

At Davidson, word got around quickly: Taylor Wilson, the skinny, excitable kid with the crazy Southern accent, had created his own star. At first, few knew what to make of it.

"I was like, this is a metaphor, right, Taylor?" Ikya said.

"Not really," Taylor said. "Stars are powered by nuclear fusion, and that's what I made. It was hotter than the sun in there." Words started to fly around the school—*nuclear fusion, neutrons, ions, radiation*—and Harsin fielded more calls from worried parents. Between the parades of students, professors, and administrators from Davidson and UNR who came through the lab to see his machine, Taylor continued to tweak it, experimenting with ways to increase its neutron output.

Over the next few weeks, he started using the fusor to irradiate different materials, bombarding them with neutrons to see how they would react. "I activated fifteen different elements with radiation," he excitedly told Walenta, "and made atoms that had never existed on earth before."

Taylor could have taken the fusor to the Western Nevada Regional Science Fair as it was and won. But he had higher ambitions. The fusor was a promising neutron source for research. But a conventional fusor emits too few neutrons for some applications, including Taylor's medical-isotopes project. Taylor had an idea: He wanted to see if he could increase the neutron output by incorporating fissionable isotopes of heavy elements, using the neutrons from the fusion reactions to induce fission in the atoms of the heavy elements, which would produce more neutrons.

Again, the progression of his experimentation was following roughly the same path that particle physics itself had followed: from the discovery of x-rays and radioactive compounds to the use of alpha particles to induce nuclear reactions to the development of electrostatic accelerators and artificial radioactivity to, finally, nuclear fission and fusion.

As the science fair approached, Taylor experimented with different amounts and combinations of thorium, natural uranium, and depleted uranium compounds and metals. The data showed that the multiplication effect he'd predicted was indeed taking place and that it increased as he upped the voltages.

"That was actually quite significant," explains Phaneuf. "He systematically developed that concept of multiplication by starting with a fusor and combining technologies, which was a novel idea."

Taylor had found a way to multiply the neutron output of a fusor. And he had found the focus of his first science fair project, which he called Subcritical Neutron Multiplication in a 2.5 MeV Neutron Flux.

But the week after he came up with his project's title, Taylor came home from the lab with his shoulders slumping.

"What's wrong, Tay?" Tiffany asked.

"We got shut down," Taylor said. "The university shut us down."

"Safety concerns," Taylor told his parents and Ikya and Sofia. "That's all they said, safety concerns. Beyond that, I don't know anything—just that I can't go in there and we can't do any more work on it."

With Phaneuf out of town, Brinsmead called an emergency meeting at the Hub coffee shop, though when Taylor got there, Brinsmead had yet to arrive. He sat down at an outdoor table near the street with physics department technicians Wade Cline and Andrew Oxner. Cline and Oxner sipped coffee; Taylor drank hot chocolate.

"I think it was Winterberg," Taylor said. The eccentric German who had made no secret of his initial opposition to the project had followed up with regular and repeated warnings during faculty meetings that it was a bad idea to have a child anywhere near the radiation-producing machines in the department's basement, let alone building a reactor that would produce lethal levels of x-rays and neutrons. But with Phaneuf serving as chair of the university's radiation safety committee, and with the rest of the faculty quickly becoming an enthusiastic fan club for Taylor, Winterberg's protests had been acknowledged and noted, but never acted on.

"I think he's crazy," Taylor said, getting keyed up on sugar and indignation. "Basically, he's just a crank."

"I wouldn't consider him a crank," said Cline. "Although he *is* cranky, and he will argue with anything and everything. Probably his biggest contribution to the department is scaring off prospective physics students."

"I'm not so sure it's Winterberg," said Oxner. "I haven't seen him downstairs in a while. He's theoretical only. If he gets into anything applied, he gets into trouble. Remember the time he was using a Mixmaster to show that mechanical energy could be turned into heat? That's why he stays with theory now."

As they talked, a 1973 Cadillac hearse slipped down the street, soundless as a ghost, and parked in the no-parking zone directly in front of the café. Brinsmead opened the car's door and was immediately accosted by two guys leaving the café who wanted to know all about his electric Caddy. "I bought a bunch of buses in Santa Barbara and towed them to Reno to get the NiCad batteries out and put 'em back there where the casket would go," Brinsmead said. "It cost more to build than it cost to buy my Nissan Leaf. But it only costs ten cents a mile to run it."

Brinsmead finished showing off his hearse, then walked toward the table, pulling his phone out of his pocket. "It's my mom," he said. "She's ninety-four, and I got her a phone. Now she's fallen in love with texting." Brinsmead sat at the table, and within moments his coffee appeared (all the other customers had to fetch theirs at the counter).

"Ron's down at Lawrence Berkeley and I haven't been able to reach him, but I don't think it was Winterberg," Brinsmead said, responding to Taylor's earlier texts. "Although you're right, Taylor, he *is* crazy, and eccentric as hell . . . or maybe just senile. He's always suing someone, for age discrimination or something. The last one was for 'stealing' his ideas — but you can't patent gravity, if you know what I mean."

So who shut down Taylor's fusion project?

"I think it was a gang of university administrators," Brinsmead said, naming three people who, he said, were irked by the fact that Taylor's high-profile project was bringing recognition to the wrong people — in particular, to Bruno Bauer, "who they're always trying to sabotage.

"I heard they came in after hours," Brinsmead reported, "with some sort of trumped-up fake safety inspection. They knew Ron wasn't in town and nobody was there. There was a gaggle of about six and they must have found some things to nitpick about."

"I understand why Bill thought that," Phaneuf tells me later, "because he's been on the wrong side of a lot of some very dubious politics. But most of the people in the administration supported the project. Of course, you're always going to have some people who are enforcers rather than enablers, so you learn to avoid them. But you can't always."

• • •

When Phaneuf got back to town, he found out that university administrators had halted the project because of something that had happened a week earlier.

"Ah yes," Taylor would later say, "the infamous pulser incident."

In their quest to multiply the neutron output of the fusor, Taylor and Brinsmead continued to bring up the power. That put stress on the fusor's reaction chamber, which incorporated a cooling jacket that they hadn't yet needed to use. When it got hot enough to cause the refractory metal to glow, they installed and tested the cooling system. But the neutron output was still limited by the amount of power coming out of the wall plug. In an effort to boost the peak voltage and increase neutron output, Taylor got ahold of a capacitor called a pulser that could store electric current as it charged over time and then discharge it very rapidly in ultra-high-voltage bursts.

Phaneuf was skeptical. "You'll get a larger burst," he said, but by pulsing every ten or twenty seconds in intense bursts, the average power won't change, and so the total number of fusion reactions and resulting neutrons will work out to pretty much the same. "Plus, you're going to put a lot of stress on your materials," he told Taylor.

Though Phaneuf was dubious that pulsing the fusor would result in any additional neutron multiplication, he thought it wouldn't hurt for Taylor to try, if only to prove the point. Once Taylor hooked the pulser up to the fusor, Taylor and Brinsmead were able to greatly increase the peak voltages. They started making test runs, bringing the pulses progressively higher, to fifty thousand volts and beyond.

One day, they pushed the voltage up further than they'd ever gone before. With the increased fluxes there was more fusion and consequently more fission; the instruments were picking up much higher neutron output from the surrounding materials. Taylor urged Brinsmead to let him nudge it up another notch.

"I don't know, Taylor," Brinsmead said, sniffing as he picked up the telltale kitchen smell of overheated stainless steel, "we're really hammering it. That pulser's really well insulated, but if we push it too far . . ."

Suddenly, the pulser's capacitors started arcing.

"This is getting a little scary," Taylor said.

"Let's back it down!" Brinsmead yelled. "I'm worried the grid might blow up."

As Taylor was reaching for the voltage dial there was a bang; it sounded like a rifle shot. Up and down the hallway, panicked professors and postdocs came running out of offices and laboratories. Among the first to arrive was physicist Jonathan Weinstein, who worked in the lab next door.

"He ran in completely freaked out," Brinsmead says, "and he said that one of his lasers had just fired."

In addition to blowing a hole in the pulser's large, expensive ceramic insulator, Taylor and Brinsmead had overloaded the physics lab's electrical circuits and caused Weinstein's laser to spontaneously fire.

Weinstein was afraid that, in addition to causing electrical safety issues, Taylor and Brinsmead could be irradiating the department—which wasn't irrational, given that the laboratories were separated only by a Sheetrock wall.

University officials halted the project to allow a thorough assessment of the situation. Meetings were called; inspections were ordered. Administrators and the university's radiation safety officer came through the lab with radiation detectors and checked everyone's dosimeter.

They found no signs of dangerous radiation exposure. Taylor's growing cadre of supporters in the physics department, including Weinstein, whose laser had fired, were in favor of letting Taylor continue with his project. Nevertheless, the UNR administration remained skittish. It wouldn't be good for the university, or anyone, if a fourteen-year-old boy was fried in the physics lab. At a meeting with Davidson Academy Director Harsin, university officials expressed their reluctance to let the project go on—even though, as Harsin made clear, the science fair was quickly approaching.

After the meeting, Harsin got on the phone with Kenneth, and Bob Davidson. Davidson drove down from Lake Tahoe and met with the university administrators. He listened, heard everyone's concerns, and asked questions.

And then he walked over to the university president's office.

Within a couple of days, Taylor was back in the lab preparing for the Western Nevada Regional Science Fair. Bringing a nuclear reactor to a regional science fair was, as Judy Dutton put it in her book *Science Fair Season*, "like bringing a Ferrari to a go-cart race." Taylor's first-place prize qualified him for the Intel International Science and Engineering Fair (ISEF), the largest precollege science event in the world.

Coincidentally, the science fair was in Reno that year—but there would be no hometown advantage. ISEF is the Super Bowl of youth science endeavors. Taylor would be competing against fifteen hundred über-geniuses from fifty countries.

Brinsmead and Taylor put the fusor in the back of Brinsmead's electric van and brought it over to the convention center.

There were no baking-soda volcanoes or potato clocks in the hall. Strolling through the aisles, they saw a kid with five patents and twelve million dollars in venture capital presenting a nanotechnology project; a kid who had come up with a new way to clean up oil spills; a kid who had developed an anti-tsunami device to break up waves before they reached the shore; a girl who had genetically engineered "smart worms."

George Ochs had coached Taylor on how to prepare his supporting materials and how to interact with the judges—who, he said, would question Taylor rigorously about his project. "At first, he'd ramble, and he wouldn't get to the point about the project," says Ochs. "But we practiced and role-played, and he got pretty good at answering concisely."

Taylor's fusor was a big hit, but his presentation materials were a bit rough, which cost him. Still, his project came in fourth in Physics and Astronomy and second in Vacuum Technology. His invention also caught the attention of Intel CEO Paul Otellini. Hearing that a kid had brought a nuclear reactor to the fair, Otellini grabbed a colleague by the sleeve and made a beeline for Taylor's exhibit.

"The first thing he wanted to know," says Taylor, "was how a fourteen-year-old could get his hands on all the stuff to build a nuclear reactor."

Taylor explained everything. A few minutes later, Otellini was seen walking away open-mouthed and shaking his head in what looked like disbelief. Later, I would ask him what he had been thinking.

"All I could think was, *I am so glad he is on our side.*"

PART V

25

A FIELD OF DREAMS, AN EPIPHANY IN A BOX

THE MORNING AFTER our Bayo Canyon expedition near Los Alomos, we pick up Willis at his hothouse for a final day of atomic adventuring. Willis directs Tiffany past the Albuquerque International Sunport (the airport) then along a dirt track that crosses a mesa south of the runways shared by the civilian airport and Kirtland Air Force Base.

The U.S. Food and Drug Administration had cracked down on radioactive quack cures before World War II, but after the war, a new kind of nuclear miracle-mongering sprang up, this time with the government playing lead salesman and dream merchant for atomic schemes that promised to do anything and everything. There was Project Plowshare, which proposed using nuclear explosions to excavate a sea-level waterway through Central America (nicknamed the Pan-Atomic Canal). There was Gas Buggy, which envisioned a sort of nuclear fracking to coax natural gas out of the ground. Project Chariot would have used hydrogen bombs to create a harbor at Cape Thompson, Alaska. The radioactive fallout from Project Chariot's proof-of-concept test, which blew a massive crater in the Nevada desert, contaminated more Americans than any other nuclear event.

Spurred on by patriotic PR campaigns, big boys with big bombs and big budgets used the desert landscapes of the Southwest as their experimental playgrounds. As the strategy of nuclear deterrence catalyzed a frightening, paranoia-driven arms race, the bombs got bigger and more numerous, and the number of Broken Arrows—incidents in which nuclear weapons were accidentally lost—began to grow. A lost nuke is unsettling enough; even more unsettling is the fact that it happened so often that the armed forces actually needed to coin a term for it. By the military's own admission, there have been at least thirty-two Broken Arrows since 1957 and many other "oops" incidents with extreme-calamity potential. (As recently as 2007, nuclear cruise missiles were accidentally loaded on a B-52 and flown across the country.)

Taylor, fascinated by some of the more bizarre and little-known events of the Cold War era, began filing Freedom of Information Act requests. He learned that in 1958, off the coast of Georgia, a B-47 carrying a 7,600-pound hydrogen bomb collided with another plane in midair, and the bomb was jettisoned. It sank into the ocean and was never found. In 1965, an A-4 Skyhawk jet rolled off the side of the USS *Ticonderoga* and sank, with its hydrogen bomb, sixteen thousand feet down into the Pacific Ocean; it too was never recovered. In 1968, a B-52 carrying four nukes contaminated a large strip of sea ice and snow when it crashed near Thule, Greenland. The seven-hundred-person cleanup operation was nicknamed Dr. Freezelove.

Considering the lack of consistent safety standards during the early years of the arms race, it now seems remarkable that there was never an inadvertent nuclear detonation. According to the military, only two accidents resulted in widespread dispersal of nuclear materials. But as Taylor would discover, this was either wishful thinking on the military's part or an intentional fib.

One night, as Taylor was reading through a partially redacted Pentagon accident report, he came across an incident that stood out for its disturbing similarity to the absurdist fiction of *Dr. Strangelove.* What also stood out was its location. Taylor immediately picked up the phone. "Carl!" he said. "There's a Broken Arrow right in your backyard!"

On May 27, 1957, a ground crew at Biggs Air Force Base in Texas loaded a bomb into the bay of a B-36 Peacemaker bomber headed for Kirtland Air Force Base in Albuquerque. The Mark 17 was the largest bomb the United States ever made, seven hundred times more powerful than the Hiroshima bomb and so heavy that the B-36 was the only aircraft capable of lifting it.

As the plane made its final approach to Albuquerque, a crew member went back to check the bomb's locking pin. To this day, it's not exactly clear what went wrong (the last paragraph of the Pentagon document was redacted), but the airman most likely pulled a lever in the wrong direction, causing the forty-ton bomb to drop from its mounting slings and smash through the bomb-bay doors.

The crewman lunged for a handhold and managed to avoid his own Slim Pickens moment as the Mark 17 plunged seventeen hundred feet down to the mesa. Though the bomb's plutonium core hadn't been inserted, it contained a spark plug made of conventional explosives and either plutonium or enriched uranium. The conventional explosives detonated on impact, taking out a grazing steer and creating a fireball that was seen and heard throughout Albuquerque.

Tiffany parks the SUV among the mesquites. Willis has visited the site several times now, and Taylor has been here twice before. We unload metal detectors and Geiger counters, and Tiffany grabs her coffee cup. Then we fan out across the mesa.

"This," Tiffany says, smiling, "is how we spend our vacations."

According to the cleanup report, the crater was filled and the contamination completely removed. "You kinda gotta read between the lines," Taylor says. "*Completely removed* means something like 'We bulldozed most of the stuff into the crater, covered it, then took a cursory look around.' *Unrecovered* means they decided it wasn't worth the effort. There's an unrecovered reentry nose cone from an ICBM that went off course and burrowed into the ground somewhere in New Mexico. Carl," Taylor says. "We should definitely go there."

We search near ground zero and find nothing. Then we spread out to the north and east. After nearly an hour, our detectors begin to beep. We find bits of charred white plastic and chunks of aluminum — one of which is slightly radioactive. These are fragments of the hydrogen bomb. I uncover a broken flange with the screws still attached, and Willis digs up a hunk of hot lead.

"Got a nice shard here," Taylor yells, finding a gnarled piece of metal. He scans it with his detector. "Unfortunately, it's not radioactive."

"That's the kind I like," Tiffany says, smiling.

We keep walking east, closer to Sandia National Laboratories Area 5, with its Annular Core Research Reactor and Plasma Materials Test Facility. "You can't see it," Willis says, "but over there is Kirtland's underground munitions storage area. There are more nukes there than anywhere else on earth."

We watch a Black Hawk helicopter leave the base and head south. Abruptly, it diverts from its course and starts toward us. "They've spotted us," Willis says. "Happened last time too. You always wonder if they're drawing a bead on you with that M134 Minigun."

"Six barrels and six thousand rounds a minute," Taylor says, looking up as the copter circles us, then continues on its way. Taylor turns on the metal detector, which starts beeping immediately. He digs with his hands, pulling up small pieces, many radioactive. "Oh my God, Carl, it's hot out here!" he yells. "This is so exciting! This place is loaded!"

Suddenly we're finding radioactive debris under the surface every five or six feet—even though the military claimed that the site was completely cleaned up. Taylor gets down on his hands and knees, digging, laughing, calling out his discoveries.

Willis is having fun too, and so am I. "It's the thrill of the hunt!" Willis says. I find something that looks like it's made of Bakelite, an early form of plastic. Then Willis picks up the coolest thing we've seen yet, a large chunk of the bomb's outer casing, still dull green. He calls Taylor over.

"Wow, look at that warp profile!" Taylor says, easing his scintillation detector up to it. The instrument roars its approval. Willis, seeing Taylor ogling the treasure, presents it to him. Taylor is ecstatic.

"It's a field of dreams!" he yells.

Tiffany checks her watch. "Tay, we really gotta get going or we'll miss our flight home."

"I'm not even close to being done!" he says, going back to digging. "This is the best day of my life!"

"Taylor just hates for things to end," Tiffany says, watching him with exasperation and amusement. "He likes to drag everything out till the last minute."

By the time we manage to get Taylor into the car, we're running seriously late—and we've got a trunk full of uranium ore, bomb fragments, and shards that we've collected over the past few days.

"Tay," Tiffany says, "what are we going to do with all this stuff?"

As we race toward Willis's house and then the airport, Tiffany's question becomes more critical. "Well . . ." Taylor says, dragging out the word like southern Arkansas folks do when they're working out what to do about something but not in any kind of expedited way. Tiffany, at this point, is passing through bemusement toward annoyance—a state in which I've rarely seen her.

"Carl," she says, "want to keep it at your house and we can give you some money to ship it back to Reno?"

Willis isn't too interested in adding another item to his to-do list. "I think the more of this stuff we can get out today, the happier we'll all be," he says. "Listen, for fifty bucks, you can check it on as excess baggage. You don't label it, nobody knows what it is, and it won't hurt anybody."

A few minutes later, we're folding closed the flaps of a too-flimsy box and taping it shut. As we load it back into the trunk, Taylor says, "Let's see, we've got about sixty pounds of uranium and a bunch of bomb shards and radioactive shrapnel. This thing would make a real good dirty bomb."

Taylor is again exaggerating. In truth, the radiation levels are low enough that, as long as no one stands too close to it for an extended period, the cargo poses little danger. Still, we stifle the jokes as we pull up to the curbside check-in. "Think it will get through security?" Tiffany asks Taylor.

"There aren't radiation detectors in most airports," Taylor says. "In fact, the lack of radiation detectors in our transportation system is, in my opinion, a pretty major security problem."

While the skycap weighs the box, I scan the Transportation Security Administration (TSA) prohibited-items sign. Spray paints, fireworks, and bleach can't be checked on, but radioactive materials are not specifically mentioned. That seems odd, considering that possibly the only thing Barack Obama and Dick Cheney ever publicly agreed on was that their worst fear was a nuclear terrorist attack on American soil. (Later, a TSA spokesperson would respond to a query about this with a clarification of sorts: "When a hazmat is not listed in the exceptions it does *not* mean that it is allowed in baggage.")

The skycap drops the box onto the belt and takes Tiffany's credit card.

"Taylor," she says, turning to her son, "you owe me fifty bucks."

"No problem," he tells her. "I can sell one of those uranium rocks on eBay and make a hundred dollars."

When we go through security, Taylor gets pulled aside so the TSA officers can take a look at the gadgets in his carry-on bag. The Geiger counters and metal detectors attract quite a crowd of officials, but the real showstopper is Taylor's belt-mounted neutron ray radiation detector. "It can detect gamma rays and neutrons," the boy tells the circle of agents. "You wouldn't believe the kinds of things I've found with it."

The agents call for their supervisor, who asks Taylor what he's doing with all this gear.

"I'm an amateur applied nuclear physicist," Taylor says.

The supervisor turns the detector over in his hands. "Wow," he says, "that's cool."

"Hey!" Taylor says, catching sight of a whole-body imaging device that's just being installed (they're now common in airports nationwide). "You guys got a passive millimeter wave scanner—I know the guy who developed that!" He leads several TSA guys over to the device and gives them a walk-around explanation of the scanner and its features. "The amazing thing about these," he says, "is that they don't actually hit you with any radiation; it uses radiation coming off your body and whatever you're wearing to create the images."

A couple of hours later we land in the Reno airport and head for the baggage-claim area, where Kenneth and Joey are waiting. We scan the carousel for the box.

"Hope it held up," Taylor says. "And if it didn't, I hope they give us back the radioactive goodies scattered all over the airplane."

Soon the box slides onto the belt and makes its way slowly toward us. Joey moves to intercept it.

"Better let Dad," Taylor says. "That stuff's pretty hot."

Kenneth picks up the box, which is adorned with a bright white strip of TSA tape. Inside, there's a note explaining that the package had been opened and inspected by the Transportation Security Administration.

"They had no idea," Taylor says, "what they were looking at."

Taylor's airport-security experience led to his second fusion epiphany. He'd been thinking about what kind of project he could do for the next International Science and Engineering Fair (ISEF), which would be in San Jose. To truly compete against the planet's brightest young scientists, Ochs had told him, he'd need to demonstrate a compelling real-world application. His medical-isotopes application was the obvious choice, but he was still a long way from the proof-of-concept phase.

Over the previous several months, Taylor had gotten increasingly consumed by issues of nuclear proliferation and nuclear terrorism—so much so that he'd started having nightmares about being kidnapped by hostile agents who milked him for knowledge.

According to the *New York Times,* "The world is awash in 2,000 metric tons of weapons-usable nuclear material spread across hundreds of sites around the globe." The International Atomic Energy Agency says that about 100 incidents of theft and other unauthorized activities involving nuclear and radioactive material are reported every year.

Although it's theoretically possible that terrorists or rogue governments could assemble enough weapons-grade fissile material to build a crude nuclear bomb, experts say a far more likely threat is a dirty bomb, in which radioactive materials are dispersed with conventional explosives. According to George Moore, a former senior IAEA analyst, "Many experts believe it's only a matter of time before a dirty bomb or another type of radioactive dispersal device is used."

Taylor came across a report about how the thousands of tons of air and ocean cargo entering the country daily had become the nation's most vulnerable "soft belly"—the easiest entry point for weapons of mass destruction. Security experts have been struggling to devise ways to strengthen cargo security without paralyzing global trade. Most cargo is packed into shipping containers, which are unloaded from container ships and dropped onto waiting trucks and railcars. Only a small percentage of the containers are checked, a process that is currently done manually.

Taylor knew, from his past few months of experiments, that he could use neutrons from a fusion reaction to force heavy atoms into fission. Lying in bed one night, he hit on an idea: Why not use a fusion reactor to produce weapons-sniffing neutrons that could scan the contents of cargo containers as they passed through ports?

Over the next few weeks, he devised a concept for a drive-through device that would use a small reactor to bombard passing containers with neutrons. If fissionable weapons materials were inside, the neutrons would induce the atoms into fission, emitting gamma radiation. A detector, mounted opposite, would pick up the signature and alert the operator. While working through the design, Taylor had a flashback to his nuclear chemistry class and realized that the neutrons would, if they encountered conventional explosives, activate nitrogen nuclei that would emit detectable gamma rays.

"Using neutrons as an interrogator made sense," says Phaneuf. "And no one had come up with a way to do it using a small, portable device, so it was quite an innovative and timely idea."

Taylor named his science-fair project Fission Vision: The Detection of Prompt and Delayed Induced Fission Gamma Radiation, and the Application to the Detection of Proliferated Nuclear Materials.

It was one of those "Why didn't anyone ever think of this before?" ideas, as evidenced by the buzz around Taylor's exhibit at the International

Science and Engineering Fair, where Taylor won four awards, thirty-five hundred dollars in cash, and an expenses-paid trip to CERN, the European Organization for Nuclear Research, in Geneva, Switzerland.

At ISEF, tech companies are always in attendance, scouting for talent. So are representatives from top universities and research institutions such as the National Institutes of Health and the Department of Energy, all of them looking to tap into the prodigious talent in the exhibit hall. Rumors abound that operatives from spy agencies such as the CIA, the FBI, and the NSA (and several foreign governments) are also milling about, although the CIA is the only agency that would comment. When asked, CIA media spokesman Edward Price said agents from the CIA's Directorate of Science and Technology, which is devoted to recruiting talent, especially physicists and mathematicians, attend science events and fairs, "but I'm not able to speak to whether we've participated in the specific venue you mention."

The Department of Homeland Security wouldn't confirm whether its agents were there, but shortly after the fair, the DHS invited Taylor to Washington to meet with officials from the Domestic Nuclear Detection Office.

"It was hard to describe how crazy it felt," says Tiffany, who accompanied Taylor, "to see him sitting at a conference table with these high-level officials treating him as an adult, and looking to him for answers. I realized then that he wasn't just my boy, he was someone who was really going to make a difference."

All agreed that container traffic was a vulnerability that needed to be addressed, and they talked about how Taylor's inventions could be put to work to intercept weapons that terrorists might try to smuggle into the country. DHS officials invited Taylor to submit a grant proposal to develop the weapons-detector design.

While in Washington, Taylor also met with Undersecretary of Energy Kristina Johnson, who says the encounter left her "stunned."

"I would say someone like him comes along maybe once in a generation," said Johnson, who also participated in science fairs in high school. "He's not just smart; he's cool and articulate and unbelievably focused. I think he may be the most amazing kid I've ever met."

The trips to Washington and Switzerland inspired Taylor, but not as much as the ISEF experience itself had. Building his reactor, conducting and documenting his experiments, and presenting his work to the judges had all helped to build his confidence. And watching the winners at the awards ceremony had opened his eyes to bigger possibilities.

One day, he rushed into Phaneuf's laboratory. "Ron, I've got an idea!" he said. "And I think it may be the coolest thing I've thought of so far."

Taylor had been brainstorming ways to complement his active-interrogation system—which used nuclear fusion to detect weapons-grade uranium and conventional explosives—with a weapons-grade plutonium detector. Plutonium emits radiation in the form of neutrons and gamma rays, but the emissions are not difficult to shield, making it all too easy to transport small amounts of plutonium without detection. Neutrons don't carry a charge and are impossible to detect directly, but they can create reactions in absorber materials that subsequently create signals for detection. Security agencies typically use neutron detectors that incorporate an inner layer of helium-3, a lightweight isotope of the gas that lifts birthday balloons; it reacts with plutonium's emitted neutrons to form charged particles that register in the detector. But helium-3 is the rarest gas on Earth, and supplies are quickly running out. Though U.S. security agencies have sought to install neutron detectors at ports and border crossings since 2001, the short supply and expense of He-3 (it sells for several thousands of dollars per liter) has prevented widespread deployment of plutonium detectors.

Taylor realized, he breathlessly told Phaneuf, that he might be able to replace the He-3 detectors with a portable neutron detector made with the cheapest, most available substance on Earth: water. His brainstorm came after he traveled to CERN, where he'd gotten a firsthand look at the large dielectric detectors that astrophysicists and particle physicists use to detect rare particles like neutrinos and muons. The detectors make use of the so-called Cherenkov effect, named after the Russian scientist who discovered that subatomic particles can travel through a medium such as water faster than light can travel in that medium. When this happens, the particles produce light (the brilliant blue glow of an underwater nuclear reactor core), which can be measured.

"Of all the ideas Taylor had come up with," Phaneuf says, "this was the most novel so far, and the one with the most commercial potential. Cherenkov detectors have been used in astrophysics and particle physics for many years, but they'd never been used in a portable neutron detector and applied to counterterrorism."

Taylor went to work on the project in Phaneuf's lab, experimenting with different materials and configurations. He settled on gadolinium (a rare-earth mineral known for its exceptionally high absorption of neutrons) as a doping compound and built a scaled-down, portable prototype. After a few months of trying different concentrations and forms of gadolinium ("the secret sauce," Taylor jokes), he used his fusor to generate neutrons to test the device.

Thus, the first portable water-based neutron detector was born. Significantly, his detector was more sensitive than existing helium-3 detectors and could be manufactured for roughly a thousandth of their cost. Taylor now had a passive method for detecting plutonium, and he combined it with his active Fission Vision systems to create a comprehensive suite of counterterrorism detection devices. He called his ISEF entry Countering Nuclear Terrorism: Novel Active and Passive Techniques for Detecting Nuclear Threats.

That year, Ochs again worked with Taylor to prepare for the fair, to be held in Los Angeles. "Okay," Ochs said, "your first year you didn't have enough data, and your display was minimal, a little cartoonish. You can interview well, but you need data and facts to tie in."

"I think data collection and documentation is the biggest area I've grown in," Taylor said. "I've gotten more precise about it."

"You're right," Ochs said. "Your display is a lot better, more polished. This year, I want to work with you about listening and thinking before you speak, so that your elevator speech doesn't turn into a PhD dissertation."

"Going into the fair, I thought I had a pretty good entry," Taylor says. "But I didn't know how the judges would react. Then I started noticing there were more judges than usual coming by."

Before long, Taylor's exhibit was swarmed. Among the visitors was Harvard professor Dudley Herschbach, who won the Nobel Prize in Chemistry in 1986. Herschbach, who overcame intense skepticism about what he calls his "lunatic fringe" research using crossed molecular beams, immediately hit it off with Taylor.

"I could tell he had a contagious enthusiasm for knowledge," Herschbach says. "Just like me."

Shortly after Herschbach departed, Taylor was approached by Renée Montagne of NPR's *Morning Edition*. She switched on her recorder.

"Taylor Wilson, you're fifteen, sixteen years old?"

"Just turned seventeen Saturday," Taylor replied.

"You know, what would you say is the motivating force for you?"

"So, you know, I've been doing applied nuclear physics for, I would say, six years now. And there's some people at the science fair, for example, in the physics category, that enjoy working on, you know, solving the questions and the mysteries of the universe, you know. How did we get here? Where did we come from? Me, on the other hand, I like taking new things that've been discovered and applying them to real-world problems. So most of the research I do is trying to solve a problem, whether it be terrorists bringing nuclear weapons through ports or curing cancer with radioactive material. So it's taking this new physics that's been discovered and applying it."

Montagne mentioned that many of the entrants had patents and asked Taylor if he had any.

"Yeah," Taylor said. "One patent pending, and then three other patents in the process."

"Okay. Your patent pending?"

"It is a new neutron detector for the Department of Homeland Security. The current material used for neutron detection is the rarest and most expensive substance on planet Earth. It's called helium-3. I've developed a neutron detector that uses water. So we're going from the most expensive, rare substance on planet Earth to the cheapest and most abundant."

The awards ceremony was at noon on Friday. While Tiffany, Kenneth, Joey, and Ashlee sat in the middle of the crowd, Taylor sat with Ochs and the rest of the Nevada team as the celebrity emcee presented the awards for the various categories. He finally arrived at Physics and Astronomy.

"I had an inkling I might win an award, maybe third or even second place, because of all the buzz around my display," Taylor remembers. In fact, his project was one of the first-place winners. Taylor trotted up to the stage to accept the award and was told to wait. In a moment, he found out why: he'd also won the Best in Category award for Physics and Astronomy, which came with a five-thousand-dollar cash prize. After receiving congratulations, he went to pick up his award. Backstage, the other Best in Category award winners were lingering; Taylor joined them. As they exchanged congratulations, one of the organizers approached him.

"She told me I needed to pick up my award, then go back and sit with my team," Taylor says.

Something was up.

All that was left were the Gordon E. Moore Award and the two Intel Foundation Young Scientist Awards. The emcee relinquished the stage to the Intel Foundation's executive director, Wendy Hawkins, who said that the Young Scientist Award winners were selected for their commitment to innovation in tackling challenging scientific questions, using authentic research practices, and creating solutions to the problems of tomorrow. The award, which seemed so far out of reach to most of the finalists that they didn't even dare to imagine winning it, came with a fifty-thousand-dollar cash prize.

"Taylor," Ochs whispered as Hawkins wound up her speech, "I think you've got a chance."

The first winner was a team from Thailand.

"Our other winner of the Intel Foundation Young Scientist Award," Hawkins said, "is Taylor Ramon Wilson, from Reno, Nevada, USA!"

Taylor looked at Ochs questioningly—*Did she just say my name?*—then leaped from his seat and sprinted toward the stage. Hawkins continued: "Taylor has developed one of the lowest-dose and highest-sensitivity interrogation systems for countering nuclear terrorism."

As applause thundered and cameras flashed, Taylor ran up the stairs and was soon covered in falling confetti, hugging Hawkins, the other contestants, and anyone in reach. Then he and the other winners lined up across the riser.

"I was scanning the crowd for my mom and dad and Joey and Ashlee"—they were on their feet, screaming and hugging each other—"but the lights were too bright; I couldn't see them out there."

And so Taylor stood, smiling broadly, holding his plaque high over his head as the confetti continued to fall.

26

THE FATHER OF ALL BOMBS

"DAD, DID YOU get my coffee creamer?"

"It's out in the garage, Tay. Let me see if I can remember where I put it." As Kenneth pushes himself up off his chair, Taylor is already through the kitchen door. I follow him into the garage and collide with a blast of midsummer afternoon heat.

The Wilsons arrived in Texarkana two days earlier for their annual summer homecoming. I can immediately sense that everyone is more relaxed than in Reno, folded in among old friends and family and the familiar greenery and pace of the South. Soon — it will take a few days — all of them will find their fading accents coming back and their days taken up with friends and rituals and roles that defined them for the majority of their lives.

Taylor takes a look around the garage. "These go way back," he says, affectionately kicking a stack of traffic cones as Kenneth emerges from the house and lifts a large bag off the concrete floor. "Here it is," he says, holding the bag out to Taylor, who reaches in with both hands and pulls out a couple of one-gallon canisters of powdered nondairy coffee creamer — a year's supply for a midsize office. There are two more in the bottom of the bag.

"That gonna be enough?"

"Oh, yeah," Taylor says, looking pleased. "That ought to give us a pretty nice fireball."

Taylor glances around. "Now we just need a five-gallon bucket, some fuse, and a little black powder, and we got us everything we need to fire up the Father of All Bombs."

When I ask, Taylor says, "The FOAB? It's a Russian thing. That's what they nicknamed it, to one-up the Americans' Mother of All Bombs. The official name is the Aviation Thermobaric Bomb of Increased Power. Basically, it creates the world's largest explosion short of a nuclear weapon. It works like what I'm going to do with the coffee creamer; you get a super-high-temperature fireball and a massive shock wave. Their guy, General Rukshin, said" — here Taylor tacks on his Dixie version of a Russian accent — " 'Owl dat iss alive merely evaporates.'

"And we're gonna do it tonight," he says, "at the party!"

We gather the FOAB materials and assemble them in the backyard. Taylor is in almost constant motion, darting here and there to show me something or deal with some suddenly urgent matter that will be replaced within minutes by another even more urgent matter. When he sits, he triple- and quadruple-tasks, conversing, watching TV, checking his iPhone for the latest energy-related news, corresponding with friends and fans. "I'm getting bored," he'll say after sitting for a few minutes. "Let's go blow something up."

Kenneth and Tiffany have invited several old family friends to the party; some bring along their teenage or twentysomething sons and daughters, who are all eager to hear what Taylor is up to. Near the fridge, Taylor holds court with Susan and Louis Slimer, regaling them with interesting facts on the subject of radiation hormesis, the theory that low doses of ionizing radiation activate the body's repair mechanisms and protect against disease.

Socializing in southern Arkansas is still a genteel experience. I hear a lot of "Yes, sirs" and "No, ma'ams," and I notice that children and even young adults won't refer to an older person by first name only; it's "Miss Tiffany" or "Mr. Kenneth." But Taylor has conversed easily with adults since the second grade; he's never had a sense of being subordinate to them. Tiffany, watching Taylor from across the kitchen, tells me that at parties or on vacations, Taylor would always go toward the grownups. "He's never had adults intimidate him. We worried about that. Did he have sufficient respect for authority? Not really. But most people responded positively; they were amazed that he could come up and meet them at their level. And most kids his age didn't really want to hear about black holes and dark matter."

Jeff and Jolly Woosley want to know about Taylor's plans for college. On this subject, Taylor is vague. He'd heard from recruiters at MIT and Berkeley. He had also heard from Lee Dodds at the University of Tennessee, whom he had impressed as a ten-year-old, and Dudley Herschbach, the Nobel-winning Harvard professor. In fact, Herschbach had called Kenneth too, and suggested the possibility of Taylor skipping his undergraduate degree and coming in as a nuclear engineering graduate student.

"He said Taylor is doing research that most graduates have never thought of doing," Kenneth would tell me later, "and he'd already done all these elaborate research papers and science projects. He said what if Taylor came in as a grad student and immediately got to work on his thesis paper? He thought it was ridiculous the way some of these grad students were drawing out their dissertations four and five years. He'd always wanted to see someone come in and come out with a PhD in a year's time, and he honestly believed Taylor could be the one to do it." Taylor had gotten a similar, but firmer, invitation to go directly into the nuclear engineering PhD program at the University of California, San Diego.

But as he talked about it, there wasn't a whole lot of excitement on his face.

"Okay, Jeff, now ask him what it is you really want to know about," Louis Slimer says, initiating a round of laughter and knowing grins.

"Okay, okay," says Woosley. "So, Taylor, y'all got some fireworks for us tonight?"

"Well . . ." Taylor says, relishing the attention. He leans back against the counter, crosses his arms, lowers his head, and looks up with a mock-sheepish expression. "I just might have a little bit of entertainment for y'all."

A few minutes later I'm out on the patio with Taylor and Ellen Orr, Taylor's old friend and a schoolmate (and fellow teacher in middle school). She has kinky hair and a kooky laugh. She graduated a year early from high school in Texarkana ("I got bored, so I decided to hurry up and get out of here") and she's back in town on summer break from Centenary College. Taylor has outfitted Ellen and me with propane torches, which we're using to ignite the thin lines of gunpowder he's pouring onto the patio bricks. It's eight thirty now, just dark enough for our explosive little streaks and laughter to catch the attention of the partygoers inside.

"I do believe," Taylor says as the guests drift outside, "that it's time for us to make ourselves a little fireball."

Taylor pops the polycarbonate face shield atop his head—"No one that gets burned or blinded plans on it," he says—leaving the mask flipped up as he threads a fuse through a small hole he's drilled near the bottom of a bucket. He looks around, checking for open flames, then grabs the container of black powder.

"Okay, here's the recipe," he says, pouring a bit of powder into the five-gallon bucket. "The powder is the lifting charge, and you don't need much if you do it right. Then you cover the fuse and powder with a paper towel, to separate the lifting charge from the creamer."

Taylor glances up at the small circle of people gathered around him. "I feel like the Martha Stewart of backyard naughtiness," he says.

"Actually," he adds, pouring the creamer in, "I might-could do a book on backyard science someday. But not boring stuff; it would be stuff I like to do, like this."

Leaving the mask flipped up, he stands next to the bucket and addresses the crowd, clearly relishing the ringmaster role. "So, when you have oxygen combining with an element over a long period of time, the result is slow oxidation, like rust. But when oxygen combines with something rapidly, you have a quick energy release. What's going to happen is the black powder's going to push the sugar, which is the fuel, up into the atmosphere. When it meets the air, it'll release its energy, in the form of light and heat.

"The Russians call this the Father of All Bombs. But the way we're doing it, it should be more of a fireball than an explosion. But you never know how this humidity will affect it," he says, looking up at the sky. "So, just in case, stand back. And if you hear a big bang, duck and avoid falling shrapnel."

Ellen and I exchange anxious glances as Taylor pulls a lighter out of his pocket. Taylor flips his mask down and bends toward the fuse, and I find myself reflexively taking another step back.

"Wait a minute," Kenneth suddenly says, looking around the yard. "Where's Joey?"

Joey has, thus far, done an admirable job of remaining invisible. Apart from a few perfunctory hellos, strongly encouraged by Tiffany, he has very effectively managed to dodge human interaction. Tiffany opens the back door and yells into the house.

"Joey? Joey, come on down for the fireball!"

There's no answer.

Ellen and I, who last saw him playing a video game in the upstairs computer/exercise room, offer to fetch him. We head inside, moving past photos of the two laughing, clowning brothers in younger years. Upstairs we find Joey sitting where we last saw him, at the computer. Now, though, the screen is filled not with a game but with the Wikipedia entry for sorghum. Joey doesn't look up when we enter.

"Joey!" Ellen says. "What the heck?"

"Sorghum is good," Joey says, scrolling through a comparative chart of the nutritional content of major grains. He touches the screen with a finger. "Look at the protein content compared to the others."

"But Joey," Ellen says, standing behind him with her hands on her hips. "There's gonna be a fireball. Why are you up here Wikipedia-ing sorghum?"

Joey stares at the screen for another few seconds, then answers, still not looking up.

"I'm second best at everything," he says. "I don't know."

Taylor and Joey are both tall and thin and scary-smart. But right now, it's a stretch to believe that the kid sitting upstairs in a house in Arkansas staring at pixels describing sorghum shares fraternal genes with the showman in the backyard who's preparing to incinerate a good portion of the breathable sky.

"Joey is incredibly intelligent," Ellen would tell me later. "And he has an amazing, dry sense of humor when you can dig it out of him. But it's like he got stuck in Taylor's shadow. He's such a smart kid, but Taylor just got there first."

It wasn't always like this—and sometimes it still isn't. One night in Texarkana, we all went out to dinner at an upscale restaurant. After a week back home, everyone had loosened up. Joey was in a giddy mood; he seemed more like the kid running around in the pictures and videos that chronicle his preteen life than I'd ever seen him. Once Tiffany coaxed Joey and Taylor to look at the menu, they carried on a running gag—"I weel haf zee soufflé," they kept saying in over-the-top French accents—smirking and giggling and drawing a few disapproving stares from stuffier diners nearby. When the food came, Tiffany prodded them to eat, but the boys spent most of the time messing with each other's plates. They were just two teenage brothers having a good time together.

On the surface, both boys' early childhoods looked much like a lot of other upper-middle-class childhoods. There were the PBS kids' shows and Baby Mozart CDs. There was the organic diet (Joey stopped eating meat at age four and is still a vegetarian) and bedtime rituals that included lots of reading. "As a preschooler," Tiffany says, "Joey was always escaping and getting lost in crowds. We actually decided not to go to New York City one year, because we were sure we'd have to get a leash for him if we did."

As a youngster, Joey liked playing in creeks and drawing. Computers had established more of a foothold in the house by the time their second-born was a preschooler, and he got hooked early on the wonders of the screen. "He loved those Clifford CD-ROMs, and he would stay on them till I pulled him off," Tiffany says. Whereas Taylor has never had an interest in video games—he goes onto computers only for information—Joey is more typical of the Y Generation. He got into gaming fairly young and eventually began modding (hacking and modifying the games) in collaboration and competition with his friends—who were, in contrast to Taylor's, mostly boys.

Although Joey's mathematical talents were stifled in elementary school, he charged ahead quickly at Davidson, and by his third year he was taking advanced calculus at UNR. "That was really the best thing about Davidson, especially at first," Joey would tell me. But as his brother piled up achievements and became ever more exuberantly connected with the world, Joey became ever more introverted.

"Joey was the first one interested in science," Tiffany says. "He once said Taylor stole it from him." Joey will often help his friends with math and chemistry, and in Ms. Walenta's chemistry class he got the first perfect score on the final (which surprised his teacher, since he'd refused to attend the pretest review). "But now he says he's not going to college," Tiffany tells me. "He says what's the point?"

"This is a family dynamic that I see too often," says Dona Matthews, a researcher in child and adolescent development at the University of Toronto. "When you get a kid like Taylor who's on fire, and the family pours all their resources into making sure he gets all the stimulation he wants and needs, sometimes not everyone in the family gets a chance to develop their interests and abilities."

Matthews took on these issues in her recent book *Beyond Intelligence: Secrets for Raising Happily Productive Kids*. "I don't necessarily celebrate when I hear about amazing intellectual achievement at a young age," she says. "I worry; I want to know if all members in the family are thriving, not just the superstar."

David Feldman's research for *Nature's Gambit* left him with a dire view of the outlook for siblings of prodigies. "At a minimum it's problematic, and potentially really sad for the second child," says Feldman. "In most cases what it takes to fulfill the potential of one of these kids is more than even a really dedicated family can provide. And so the second child gets the scraps."

But now, he adds, "I'm a little less sure that's true. Certain strategies can protect siblings, and I've come across examples of high-performing siblings." Feldman cites the Williams sisters in tennis and the Polgár sisters, a trio of Hungarian chess prodigies. It helps, says Feldman, if the children have very different personalities, if the sibling relationship is noncompetitive, and—most important—if there's enough parental support left over for the other children to fully pursue their interests.

"People underestimate Joey," Taylor says, "because he's not out there tooting his own horn. But I think it's simple: historically, most scientific and super-intelligent people are introverts."

That is essentially true. Among the general population, introverts are a minority, about 30 percent. But among gifted children, introverts make up between 60 and 75 percent, and the incidence of introversion goes up with IQ. In other words, among the Wilson boys' brainiac peers, Joey is the norm and Taylor—with his communication skills, his self-advocacy, his showmanship—is the exception.

"Yes, gifted children are more likely to be introverted," says psychologist Ellen Winner. "Sometimes it's because they can't find peers. Sometimes they withdraw because they don't get the attention they need to flourish in the areas they're interested in. But often it's simply because they like to spend a lot of time alone, doing what they're interested in."

Joey's shower-door calculations were a good example of the last scenario. But people who prefer quiet, alone time don't have it easy in a society that increasingly celebrates and rewards charisma and self-confident leadership. To get ahead, we're told, we must be outgoing and self-promoting.

"Those dominance-seeking behaviors seem to be the desired character traits now," says Susan Cain, author of the book *Quiet: The Power of Introverts in a World That Can't Stop Talking*. Cain says schools, which promote group learning, "can be suffering grounds and poor learning environments for introverted children, who prefer to work alone and are often reluctant to push themselves forward or make contributions without being coaxed."

Cain, an introvert herself, presents an intensely researched case (like most introverts, she excels at solitary, highly focused work) demonstrating that a large part of the population has been unfairly sidelined. "Projecting confidence and being outgoing and verbally voluble has come to be seen as the cultural ideal. Introversion," she writes, "is regarded as less successful, less desirable, and less worthy a temperament in our society."

Obviously, Tiffany and Kenneth thought Davidson Academy and the move to Reno would benefit both their children. But it was becoming clear that what worked so very well for one child wasn't working at all for the other.

Educational researchers who are focused on talent development say that being with one's intellectual peers is almost always good academically. But child psychologists have mixed emotions about segregated learning environments for gifted children. "By putting all these smart kids in the same place," says Dona Matthews, "you're creating an extremely marginal group. That kind of cloistering can intensify their differences from normal development."

Even some Davidson students feel it. "I went to London for a summer internship," says Taylor's friend Ikya Kandula, "and I realized that I'd been in an environment that sheltered me from the 'real world,' which is much more intellectually diverse. I had to figure out how to make friends."

Matthews, who grew up in London, Ontario, was pulled into a four-year gifted program during elementary and middle school. "At the time I loved it; we were doing Shakespearean plays and the teachers were brilliant. Then we were thrown back into normal high school, from the hothouse to the garden. I realized that, in terms of sexuality and dating, us gifted kids were far younger than my 'normal' sister and her 'normal' peers. They were just so much better adjusted, and they turned out to be happier as adults too."

I told Matthews that I too had been pulled into a segregated gifted program in the fifth grade—and that I'd begged my parents to let me leave it in the sixth grade so that I could return to my old friends at my old school.

"Switching back in order to be with friends," says Matthews, "is actually the best possible reason to switch back."

Matthews and others who approach the issue from a psychological (rather than purely educational) point of view say that these episodes demonstrate that each gifted child has individual needs that require individualized responses. For Taylor, the opportunity to feed his academic growth wasn't isolating at all; it made his social environment richer. Other children may fare better in the long run if at certain points social relationships are prioritized; sometimes, the academic stretch can wait.

The trick in larger schools and systems, says Dickinson-Kelley of the

Ann Arbor public schools, "is to create an environment for the gifted that doesn't make kids choose between friends and academic achievement. For young girls, especially, many if not most would give up math and science before they'd give up their friends.

"You have to think about the whole child, and work closely with the family to consider what kind of acceleration is going to work for kids who are clearly ready to take on additional challenges. We thought a lot about creating a segregated gifted program, and decided that it's better to have flexible options for accelerated learning without disrupting the social environment." That can create logistical challenges, Dickinson-Kelley says, because every kid is different.

It's not too hard to coax Joey downstairs—even introverts like to see something blow up.

"There he is!" Taylor says, looking happy to see his brother. "I was just saying this should be more of a fireball than an explosion. But," Taylor says, bending and flicking the lighter, "you never know."

I turn on my phone's crude video camera. The white bucket, reflecting the floodlight on the deck, is all that appears on the screen at first. Then the small warm flame of the lighter reflects off Taylor's facemask. He lights the fuse, then jumps up and sprints away.

"Hey, Taylor's running farther away than anyone!" Kenneth yells, laughing and stepping back. Everyone else takes another few steps back too, as the sparks creep along the foot-long fuse toward the bucket.

And then . . .

Well, what happens next is so over the top that when I show the video, even on my phone's small screen, almost everyone who watches it flinches. Then they'll say something like "Whoa!," "Oh my God!," or "Holy shit!"

I wish I'd brought better camera gear, though I doubt that even my best lens could come close to capturing the full sensory intensity of the moment. Words, too, fall short. *Massive rising fireball* comes close—but it's a very unusual fireball, an extended explosion that rises out of the bucket in a fiery vertical column so willfully that it seems to have a malevolent intent of its own. There's a noise, but as Taylor had predicted, it's not a bang—it's a freaky mega-whoosh that grows as the column rises, widens, and then blooms thirty feet above us into a tremendous mushroom-shaped ball of orange hell. Suddenly everything nearby—trees, house, spectators—is illuminated, bright beyond daylight.

Then the heat hits. Hands fly up to shield faces, and I draw in a gasp that offers nothing in the way of lung succor—there's simply no oxygen in it. My brain registers a tentative impulse of panic just as the fireball dissipates. I gulp another breath, relieved to discover that the life-giving part of the atmosphere has returned.

For a moment all is silent as everyone gratefully breathes. Then shouts and whoops fill the backyard. I look over at Joey, who has a broad smile stretched across his face, just like everyone else. Taylor approaches the bucket, flips up his mask, and gazes up at the sky that hosted his fireball moments earlier.

"Owl dat iss alive," he says, playing the Russian general again, "merely evaporates."

27

WE'RE JUST BREATHING YOUR AIR

MY POPULAR SCIENCE profile on Taylor came out just before the 2012 TED Talks. Futurist Juan Enriquez, a longtime favorite at the TED conferences, forwarded the story to Chris Anderson, TED's curator. "I read it and immediately said, 'We've got to have that kid here,'" says Anderson. A few days later, Taylor was on his way to Long Beach, where he gave a short and enthusiastically received talk titled "Yup, I Built a Nuclear Fusion Reactor."

It's not clear whether Taylor understood that an invitation to speak at TED is widely considered a career-topping signal that one has arrived as an influential thinker, communicator, and achiever. Taylor's preparation consisted of calling me a couple hours before he was due to go onstage to ask if I'd e-mail a few photographs. But his off-the-cuff talk received a standing ovation and has since been viewed online more than two and a half million times. After he spoke, several venture capitalists and other investors came out of the crowd to invite him to contact them when he was ready to fund his company.

A few weeks later, Taylor asked his English teacher, Alanna Simmons, if he could get the next week's readings in advance, because, he said, "I'm going to be out of town." Simmons asked where he was headed.

"I'm going to the White House to meet the president," Taylor said. In 2009, President Obama, as part of an effort to move the United States from the middle to the top of the pack in science and math over the next decade, had begun hosting an annual science fair to recognize the achievements of the nation's top young scientists. Obama used the occasion this time to announce that his budget proposal would include some two hundred million dollars in initiatives to support teachers and others in educating the roughly one million additional STEM graduates needed over the next decade to fill the growing number of jobs requiring these skills.

Under the gaze of the Abraham Lincoln portrait, Taylor demonstrated his fusor and his weapons-detection system and chatted confidently with the president, who joked that Taylor must have one of the most radioactive garages in the nation. While the other kids waited patiently, Taylor and Obama riffed off each other. At one point, Obama said, "Why haven't we hired you yet?"

"I kinda thought that was what I was here for," Taylor said.

"What do your parents think about all this stuff you're doing?" Obama asked.

"Well, you can ask my mom; she's right over there," Taylor said, "but basically, once they figured out that I knew how to handle it, they've given me a pretty free rein."

The president continued his rounds, stopping to shake the hands of the other students and check out their displays. On his way out of the room, he made a quick detour to Taylor and told him, as he shook his hand once more, "You may be working for me soon."

I wasn't the only one who entertained at least a brief thought that it might turn out to be the other way around. *Forbes*'s Eamonn Fingleton reported on the 2012 Halifax International Security Forum, which featured a who's who of national security notables from forty nations. "But for my money," wrote Fingleton, "the show was stolen by someone who did not figure on any of the panels – Taylor Wilson . . . a Nevada based scientific prodigy who built a nuclear fusion reactor in his parents' garage at the age of 14, Wilson held court on the fringes of the conference, confidently dispensing wisdom on everything from nuclear terrorism to the future of the world energy industry. Named the Intel Young Scientist of the Year . . . he already has a lot to be confident about and sports an extremely-young-man-in-a-hurry manner that reminds some of the early Bill Gates."

Suddenly, Taylor was overwhelmed by fan mail and media inquiries, some of which seemed, at least at first, absurd. Among them were several requests for comment on the Fukushima meltdown in Japan. Could the teenager possibly have anything relevant to say about the disaster? I wondered.

As it turned out, Taylor had begun taking local air, water, and snow samples the day after the accident. Seven days later, Fukushima's fallout arrived in Reno in the form of elevated levels of four radioactive isotopes in his sample group, which he expanded to include spinach, milk, and fruits bought at a nearby organic market.

"I've been getting lots of calls from TV producers who want to do features," he told me over the phone. Many panned out, though not always without mishap. During the taping of an extensive CNN feature, the charging resistors on a new fusion device he was building blew up. And while he was driving the crew of NBC's *Rock Center with Brian Williams* out to the Red Bluff Mine, he rolled over his dad's SUV (no one was hurt). The *Atlantic* featured him in its "Brave Thinkers" package, and *Time* magazine, in a roundup of "30 People Under 30 Changing the World," went so far as to deem him "the Next Einstein."

Suddenly, the eyes of the world were on Taylor, and he was loving it.

Though the quiet draw of scientific inquiry has long attracted introverts, science needs its extroverts too — perhaps now more than ever. Indeed, the scientific community's lack of effective communicators may be a big part of the reason why important basic research is underfunded and why science is punching below its weight when it comes to influencing policy on important issues such as climate change, vaccinations, and science education.

As the public (and the politicians it elects) becomes less science-literate, the lack of scientist-communicators has allowed many "scientific" debates to become dominated by nonscientists. Loud, zealous, and often misinformed or self-interested, these voices have become increasingly influential and immune to reasoned response.

"What happens is that as the noise from outside scientific circles gets louder, the more the sideshow gets perceived as the center ring," says NASA climatologist and climate modeler Gavin Schmidt. "Scientists need to be in that ring, correcting science that gets misrepresented. But most of us would prefer not to; most of us aren't attention-seekers."

Partly because of scientists' reticence (the media have played a role too), the gap is widening between the informed conclusions of scientists and the public's understanding. This has affected policymaking in profound ways, setting up scenarios (such as we've seen with runaway greenhouse-gas emissions and preventable pandemics) that will become increasingly problematic as their effects compound.

"I could see someone like Taylor having an important role in closing that gap," says Phaneuf, echoing the observations of many in the scientific community. "He has amazing passion, and the ability to share that passion and connect with people and explain complex subjects in layman's terms. The field needs those skills, and it needs people like him who can really popularize science itself."

When I next visit Taylor in Reno, it's clear that he's become a celebrity at Davidson. "He's this famous and really smart guy, that's how most people know him here," one of Taylor's classmates tells me. "I mean, how many people are being shadowed by these high-profile magazine journalists?"

Director Colleen Harsin says that, though Taylor's achievements have brought the school a lot of useful publicity, she's worried that he could become too one-sided: "One challenge with a kid like Taylor is, okay, you're doing a fabulous job, now you need to do other things as well." As I'm leaving Harsin's office (where, the last time we'd met, she'd warned of the pitfalls of putting high-achieving kids on pedestals), I notice that the lobby walls have been hung with framed magazine and newspaper articles featuring Taylor, the school's star student.

But in a conventional sense, as Taylor readily admits, he isn't a particularly good student. Halfway through the first semester of his senior year, Taylor is carrying a C minus in Davidson's accelerated physics class, though he is acing his upper-level physics class (radiology) at UNR. Just as he did back in Arkansas, Taylor continues to ignore schoolwork having anything to do with material that he's already learned or that doesn't interest him. Also, Taylor's expertise at gaming the teachers seems to be growing as fast as his acclaim. "We'd all give him special dispensation," Ochs, who taught at Davidson that year, would tell me later, "and he'd usually turn things in three or five days late. But almost everyone tolerated it, because . . . well, because he was Taylor."

"How are you, Mr. Wilson?" teacher Darren Ripley asks as Taylor walks into his precalculus class after a trip to New Mexico that kept him out of school for three days.

"I'm back, and that's all that matters," Taylor says, taking a seat. "There's a little bit of homework that I didn't get to, but I can do it tomorrow."

"That's fine."

"Whoops, forgot my backpack," Taylor says, hopping up. "Be right back."

"It's your world, Taylor," Ripley calls after him, smiling. "We're just here breathing your air."

In a couple of minutes, Taylor returns to the classroom and sighs loudly as he plops into his chair.

"Okay, now that we're all here," Ripley says, picking up his chalk, "shall we sally forth?"

Ripley has watched Taylor transition "from a scrawny fourteen-year-old to a scrawny seventeen-year-old who's suddenly dressing like the models in *GQ*. His peers give him a lot of flak for that, and I jump right on, because I know he can take it."

I've also noticed the changes. Taylor, when I first met him, wore hooded Hollister sweatshirts; now he's spiffy in checked shirts and skinny ties. There's something else too: During one of the network TV specials about him, he makes a not-so-casual reference to "my girlfriend."

"Oh yeah, that's big news!" he says when I ask him about it. "Shelly's a student at Davidson. But the problem is, her parents won't let her date anyone."

Not even a suitor who's hung out with the president?

"Nope, that didn't seem to make any difference. They're pretty conservative. And so I only see her at school. And once at a dance. But we write notes back and forth at night."

Unfortunately, I don't get the chance to meet Shelly.* But we do run into Taylor's good friend Ikya, whom I've met during earlier visits. Ikya starts excitedly telling Taylor a story, but Taylor gets distracted by an incoming text and interrupts her. "Okay, just a minute," he says as he types into his phone.

Ikya looks at me and sighs. "He never wants to hear about me."

That afternoon, I sit at a table in the school director's office with Harsin and Jan and Bob Davidson, who have driven down from their house in Lake Tahoe to meet with Taylor. Bob wants to introduce Taylor to a patent attorney who can help him protect his ever-expanding series of inventions.

Taylor is running late, so I bring up the worry that several of Taylor's mentors have expressed: that he might overreach and fly too high, too fast. Are the Davidsons worried about Taylor getting too far ahead of himself?

"I think Taylor may have to make a few mistakes to learn," Bob replies. "That's okay. It's very often you see these high performers having

---

* Shelly is a pseudonym; the student requested that her real name not be used.

a fine line between confidence and narcissism. But you need some of that chutzpah if you're going to make a go of it in this world."

"Y'all talking about me?" Taylor says, walking in and opening the candy jar Harsin keeps on her table. "Sorry I'm late, but it's been a huge day." He unwraps a watermelon ball and pops it into his mouth, then leans back casually on Harsin's desk. "They think the AMS"—alpha magnetic spectrometer—"telescope on the space station picked up a signal that came from dark matter. And did you hear about James Hansen's study, about how nuclear power has already saved one point eight million lives by replacing fossil fuels? And then there's the situation on the Korean Peninsula. I'm very excited!"

Jan smiles broadly as she shakes her head. "Taylor, do you ever go to classes here? How are your grades?"

"All A's," Taylor says, fibbing.

"Why don't you bring us up to speed?" Jan says. "What have you got going with your projects?"

"Okay, first, with my two front-burner ideas, I've retracted my Homeland Security proposal because I want to refine the engineering details. In addition to detecting nuclear weapons and conventional explosives, I've thought of a way to set it up to discover chemical weapons too, by using nuclear fluorescence."

Taylor says he's also been offered funding from the Department of Homeland Security and the Department of Energy to further his research into portable Cherenkov weapons detectors.

"What kind of detectors?" Jan asks.

"You're not gonna get it," Bob tells Jan. "I don't get it. The main thing is that the patent office gets it."

"And the new application?" Jan asks.

"I've got an idea for designing a specialized particle accelerator that could revolutionize the production of diagnostic pharmaceuticals, at one-thirtieth the cost and one-tenth the floor space. I've just been invited to the National Center for Nuclear Security to do proof-of-concept tests on their dense plasma focus device."

"But first," Harsin interjects, smiling, "you should probably finish your homework."

Jan Davidson looks at her husband and raises an eyebrow.

"I don't know," she says, dead serious. "Maybe not."

Then Taylor's phone rings. And Taylor, in a meeting with his wealthy benefactors, the founders of a school in which cell phones are not even supposed to be turned on, pulls his phone out of his pocket and . . . answers the call.

"Hi, Sofia," he says. "I'm in an important meeting right now. Can I call you back?"

I look at Bob Davidson, and I can see that he's looking at Taylor—not with annoyance but with admiration. He turns to me and grins, shaking his head in astonishment. *Yeah, that's what I'm talking about,* his eyes say. *Can you believe this kid?*

At the school cafeteria, Taylor has abandoned Sofia and Ikya at their usual lunch spot; he's now eating lunch with Shelly. At home, Taylor is driving everyone crazy. Plenty of teenagers go through narcissistic phases, but Taylor takes it to extremes, talking over everyone, demanding to control the TV remote, not bothering to lift a finger around the house (not that Tiffany and Kenneth have ever been sticklers for their kids doing chores). That Saturday afternoon we gather in the family living room to watch the CNN *Next List* special on Taylor, which was taped several weeks earlier.

As Taylor watches the show, he looks completely starstruck—with himself.

"He's gotten addicted to the spotlight," Phaneuf told me a couple of days earlier. "And he's started embellishing, telling the media things like he built the reactor in his garage for two hundred dollars. It wasn't built in his garage. You can't do that. And that turbo pump alone cost twelve thousand dollars, even though that's not what he paid for it. I don't know why he's doing that, but I keep telling him, 'Please don't say things like that, because people who are scientifically knowledgeable will discredit you if they hear it.'"

"Sometimes," says Dona Matthews, "by indulging that superficial showman thing, supersmart kids learn really well that if they keep up the demands, they will be met. We create a sense of entitlement and a narcissistic tendency that don't serve them well in the long run. That overconfidence, that showman thing will inevitably get challenged. If his identity is wrapped up in being a whiz kid, being anything less is going to be a blow."

Another all-too-common scenario, says Matthews, "is that the kind of energy he has been putting into science and developing just one part of himself, he's not putting into self-awareness, all of the social/emotional challenges and tasks." The adolescent years are where the consequences start to be felt, she says; sometimes these former child prodigies take a long time, even into their thirties, to catch up emotionally and socially.

"My worry," says Brinsmead, "is it could have an unhappy ending. What happens when everything he does doesn't succeed? Some burn out; they're so intense so young, then they get to be twenty and they're hanging out on Telegraph Avenue in Berkeley."

Then again, Taylor's successes were a relatively recent thing. He'd had plenty of early failures; maybe they would gird him against the downfall that so many people were worried about.

"Maybe so," Brinsmead says. "But right now he's headstrong, and he's playing games with people. It may be just a phase, and he'll get through it. But right now, I just feel sorry for his parents, and Joey."

I had already confided to Tiffany and Kenneth, who looked exhausted, that I was beginning to worry that I'd helped to create a monster.

"I think he'll get through this," Tiffany says. "But honestly, Tom, we just want it to be over. We're ready for him to move on so we can concentrate on Joey."

Taylor, as a high-school student, had already done things that would be the envy of most PhDs. Everywhere he turned, someone was calling him a genius or an Einstein. But Taylor wasn't ready to be an Einstein. And as those closest to him watched his celebrity status grow, they saw the traits that had brought him this far—his passion and curiosity, his exuberance and confidence—devolving into hubris. Taylor's arrogance and narcissism were having corrosive effects—on himself and everyone around him.

The lesson of the Icarus myth came to mind—how could it not? But Icarus's story wasn't a perfect fit. After all, Icarus hadn't put the sun in a box.

28

THE SUPER BOWL OF SCIENCE

THEN IT ALL fell apart.

"I wrote her a note every night for five months," Taylor says. "But I found out it wasn't Shelly's parents who were making the choice for her not to date me after all. It was her."

It was hard, at first, for Taylor to understand why he didn't have the right stuff. He had won the top award in high-school science. He had met, and wowed, the president of the United States and the CEO of Intel. He had a Nobel laureate begging him to come to Harvard, investors begging to fund his company, and Hollywood producers begging to make a movie about him. He'd been on TV and in the magazines and newspapers that were framed on the school's walls.

At a school for academic superstars, Taylor was the closest thing there was to a star quarterback. Why didn't she like him?

As it turned out—and this Taylor had to learn from the grapevine—Shelly was just as impressed as everyone else with Taylor's accomplishments. What turned her off was the way he was treating people.

Taylor doesn't mention that part when he tells the story to his half sister, Ashlee, and her friend Natalia over dinner one night in Beverly Hills. (Ashlee was then working at a nearby talent agency.)

"It's hard to believe that girl wouldn't go for you," Ashlee says.

She turns to Natalia. "Can you believe he's still in high school, and he's fighting cancer? And nuclear terrorism—"

"And the emergent shortage of rare-earth minerals," Taylor chimes in. "And next I'm thinking about exploring a possible cure for HIV that would shatter the virus using actinium-225, an isotope that emits alpha particles."

"Taylor," Ashlee says, setting down her fork. "Can you maybe just explain that to me without the word *alpha* in it?"

Before he has a chance to answer, Ashlee turns again to Natalia. "Can you believe this guy's my brother?"

Then, to Taylor: "Why didn't that girl want to go out with you? Was she crazy?"

"It's complicated. But anyway," Taylor says, aware of the melodrama, "I decided to drown my sorrows in science."

The HIV research would come later. Now, in his senior year in high school, Taylor was ready to take on what would prove to be his toughest challenge yet, bringing to fruition the idea that he'd had when his grandmother was dying. He would take the next step toward becoming the world-changing scientist that his eleven-year-old self had imagined. That image of his future self had inspired him and energized him. It had drawn him forward through the years, over and around obstacles, toward the solution he now saw in front of him. Taylor was ready, finally, to develop his nuclear fusion–generated medical isotopes.

In the United States each year, more than twenty million people get nuclear medicine procedures, which are part of the diagnosis or treatment of about a third of all hospital patients. Most medical isotopes are used as tracers or radioactive dyes. A tiny amount of radioactive material is injected into the body; it binds to specific tissue and emits gamma rays that are detected by special cameras, giving doctors a dynamic picture of what's going on inside bones, hearts, and other tissue.

Going back to Taylor's sixth-grade explanation of technetium and cesium, most medical isotopes have short half-lives, so they can do their job and then fade away before they damage the body. Since these delicate isotopes can't be stockpiled, doctors and patients depend on a just-in-time delivery system. Once the irradiated material comes out of a cyclotron or a reactor, such as Canada's Chalk River complex (which Taylor would tour later that year, after keynoting the Canadian Nuclear Association's annual conference), technicians work all night to process and purify batches. Then they race them, at 6:00 a.m., to waiting jets. With no time to waste, the final quality-control tests are conducted as the planes are flying to pharmaceutical companies in distant cities. Nuclear pharmacists like David Boudreaux then assemble generators and, as Taylor pointed out in his sixth-grade talk, "milk the cow." By 5:00 p.m., the finished and precisely calibrated products are delivered to medical facilities.

The use of nuclear medicine is growing around the world, but the supply of radioactive isotopes is increasingly fragile, and a looming shortage has specialists worried. The Chalk River reactor will shut down in 2016, and when it does, up to 40 percent of the world's isotope supply will vanish, with no alternative supplier geared up to fill the void. The pending shutdown has triggered flashbacks to a crisis in 2009 and 2010, when two major producers — in Canada and the Netherlands — shut down for repairs, resulting in a worldwide shortage that left patients undiagnosed and studies unfinished.

Taylor's childhood brainstorm during the dark days just before his grandmother's death had triggered his intense interest in nuclear fusion and led to so much more. Over the past six years he'd become even more convinced that nuclear fusion was the best and most efficient way to produce the isotopes. But he'd also realized that his fusor could never produce neutrons with the precision and focus needed for this application.

And so Taylor began pursuing a very different technology, known as dense plasma focus. He started to design and build a machine that could create a compressed, short-lived plasma hot and dense enough to generate a nuclear fusion reaction that could produce medical isotopes by hitting materials with five-million-volt ion beams. These sorts of energies had previously been available to medical-isotope makers only in multimillion-dollar cyclotron or linear accelerator facilities requiring large amounts of space and shielding.

He spent the next few months experimenting in Phaneuf's lab, designing, building, testing, and rebuilding. "I made three or four major iterations of the machine and dozens of minor modifications," Taylor says.

Via calculations and trial and error, he tweaked and retweaked the gas mix (deuterium, helium-3, and hydrogen) and pressure, the electrode design and layout, and the electrical network to create pulses of exactly the right shape and duration.

Taylor threw everything he had into it, consulting often with Brinsmead and Phaneuf. The project consumed him for nearly a year as he worked on one physics and engineering problem after another. In the end, he perfected a way to capitalize on the instabilities of the ultradense plasma to produce extremely energetic ion beams that could activate materials and transmute them into medical isotopes.

"The trickery," he'd told Ashlee over dinner in Los Angeles, "is pinching the plasma."

Taylor's proof-of-concept experiments demonstrated the viability of a hundred-thousand-dollar tabletop nuclear fusion device that could produce medical isotopes as precisely as the multimillion-dollar cyclotron or linear accelerator facilities could. What's more, it could produce both the short-lived positron emitters (SLPEs) used in PET imaging as well as molybdenum-99, the parent isotope of technetium-99m, the workhorse isotope in diagnostic imaging that's used in about two-thirds of all nuclear medicine procedures.

Taylor titled his ISEF entry A Novel Process for the Production of Medically Relevant Radioisotopes. It was a complex and meticulously documented project with a dense, jargon-heavy description. But behind the science, the experience was intensely emotional for Taylor. He'd finally achieved his eleven-year-old self's dream of designing a machine small enough, cheap enough, and safe enough to potentially make medical isotopes at nearly every electrified hospital on earth. But as ISEF approached, in Pittsburgh that year, he felt something he'd never felt before: a fear of failure.

As Taylor focused on his isotopes project, the senior class at Davidson was abuzz with talk of college applications, calls from recruiters, and odds-making on chances of admission to both top universities and smaller, more specialized colleges. But Taylor, who was more sought after than perhaps any other student at the academy, was feeling ever more ambivalent about college.

"I'm just not sure about what I'd get out of it," he told me. "I was giving a talk and a PhD student came up to me after and said, 'I mean, what are you going to do after all this, go to *college?*' When I thought about it, it did seem kind of absurd. Even if I was able to skip undergrad and go right into a nuclear engineering PhD program, it was hard to imagine I'd have the flexibility to do the kinds of experiments I wanted to do. And in the laboratory, I was on a roll; I was discovering some pretty exciting things on the science side of things."

Still, Taylor filled out some applications—to Harvard, Berkeley, MIT, Texas A&M, the University of Tennessee—"but honestly, I did it halfheartedly, especially the essays." Sometimes he'd even forget to have his test scores or his Davidson Academy transcripts sent.

He'd seen advanced academia, and it didn't look like he'd have a lot of leeway to do the sorts of things he wanted to. Gates and Jobs had dropped out. For ambitious entrepreneurs, maybe college just wasn't the place to be.

What Taylor really wanted to do was experiment and invent. But he wasn't sure what sort of situation would give him the flexibility to do that. Maybe he could start a company so that he could do science on his own terms and produce and sell his patented inventions. Inquiries from venture capitalists had been coming his way since he and his accomplishments came into the national spotlight.

One of those venture capitalists was Peter Thiel, who made his first fortune as a cofounder of PayPal, then made early investments in Facebook, Spotify, Yelp, and other Internet and tech high-fliers. In 2011, he began awarding Thiel Fellowships annually to twenty "uniquely brilliant" people under the age of twenty. Under the terms of the competitive fellowships, the notoriously contrarian, hyper-libertarian investor pays each fellow a hundred thousand dollars a year for two years to stay out of college.

Thiel has argued that some career paths, such as entrepreneurship, don't benefit from higher education and that the pressure young people feel to go to college is holding back innovation. He says he hopes his fellowships will encourage young people to pursue radical innovation that will benefit society. Thiel, who predicted the bursting of both the tech and the housing bubbles, says that what he calls "the education bubble" will be the next to burst. The one trillion dollars that Americans have accumulated in student debt has gone, he says, "to pay for lies that people tell about how great the education they received was."

Taylor and his parents sought advice—and were surprised that most of the educators were supportive of Taylor's pursuing a Thiel anti-education fellowship.

"A college degree is great," Bob Davidson told Kenneth. "But do you really think Taylor needs to go to college? It didn't suit Bill Gates, or Steve Jobs. Most people wouldn't imagine that going to Stanford or MIT would be a limiting experience. But it might be with Taylor. I don't want kids to be limited; I want them to fly."

George Ochs said that people like Taylor can get lost in education. "He could be at a university gathering information from others, year after year, always gaining knowledge, never using it. A Thiel Fellowship would allow him to explore all aspects of research."

But others, including Stephen Younger and Ron Phaneuf, urged Taylor not to forgo a well-rounded education. "Being exposed to a lot of different people and ideas can be a great multiplier and a growth experience," Phaneuf told him.

Whatever Taylor's choices for higher education, there was no question that he had thrived at Davidson. But increasingly, Joey was struggling. Joey had made friends at Davidson, but he still felt that his "real friends" were back in Texarkana. Academically, he'd at first excelled under Davidson Academy's agenda of letting kids fly in pursuit of their highly focused passions. What had been good for Taylor was, at first, good for Joey too. In UNR's high-level math classes, Joey could stretch his interests as far as he wanted.

But then Davidson itself began to change. Whereas the original intent was to let students pursue their passions "without putting limits on them," as Bob Davidson put it, the academy was beefing up its college-counseling services and begining to put more focus on a broader, but perhaps blander, education. The shift came partly from parents; they could see that graduating seniors who had been encouraged to pursue their chosen single-minded goals during the academy's first half-decade weren't getting into Ivy League and other top-tier universities—which was a priority for many of the more ambitious parents, especially the growing number of those who were first- or second-generation immigrants.

Davidson Academy had come into existence to swim against the very powerful stream of the dominant educational culture. But now its administrators found themselves faced with a dilemma: Should they stick with their original vision and let students pursue their narrow passions, or should they produce well-rounded students? Under the growing parental pressure, Davidson was moving toward roundedness and putting more focus on "ticking all the boxes" that college-admissions directors were looking for.

"Of course, that partially annihilates the original goal," says Bruno Bauer, the UNR physicist whose three children attended Davidson.

A much-ballyhooed accreditation process brought other changes, including less flexibility when it came to core classes such as English. Most of Joey's teachers found ways for him to make up the work he missed when an illness kept him out of school for a month, but he got little sympathy from his English instructor and eventually received a grade of Incomplete in the class.

Tiffany and Kenneth went in to talk with administrators; they advocated for their son, did what they could. But the flexibility that had been such a big part of Davidson's draw had apparently vanished—or at least been dramatically tightened up.

After Joey finished what would have been his freshman year of high school if Davidson used traditional grade levels, the family made a decision: Joey would not return to Davidson Academy the next year.

"I don't feel resentful or critical of the academy," Tiffany says, "because they're giving gifted kids an education they can't get elsewhere. Even though it wasn't a perfect fit, Joey had some wonderful math and science teachers. They've gone through some growing pains and had to make adjustments, but both our boys got a much better education over four years than they would've back home."

Just before the ISEF in Pittsburgh, Taylor flew to Hawaii for a gathering of Peter Thiel's Founders Fund. A delay on the return trip got Taylor to the convention center with only a couple of hours to spare. Taylor, jet-lagged and exhausted, quickly got his display set up.

"Now he just wants to sleep," says Ochs, who is wrangling a contingent of four Nevada students and three adult supporters, including Kenneth. But as soon as Taylor emerged from the hall, Ochs says, "there was a mob of people demanding autographs and pictures with him. Groupies! I had to put him in a hooded sweatshirt and give him a name tag that said 'Todd.'"

While Taylor sleeps, I head over to the opening ceremony. As Eurobeat music thumps, fifteen hundred kids snake through the crowded hall in packs, dancing and holding aloft their countries' flags. The air is charged with jet-lagged adrenaline; there's a light show, a ridiculously charismatic celebrity emcee, and lots of energy, pride, and hype.

During the shout-out, the emcee calls each country's name, and its contestants sprint to the stage, their flag flying. They may be the nerds back at their schools, but here, they're rock stars.

Brian David Johnson, Intel futurist, takes the stage and declares, to massive applause, "I am a geek. Geeks rock!" A minute later, Johnson says, "The most radical thing I ever did was to declare myself an optimist. The future is far too important to be negative about. The reason it will be awesome is because *we're gonna build it!*"

The next day, there's a lower-key panel of Nobel laureates and talks on financing, patents, and pirating. The ins and outs of building a business has become a popular and growing part of the ISEF program. Suddenly, it seems every young scientist wants to be an entrepreneur too.

I get a chance to walk among the displays one morning before the contestants arrive. Apart from a few standout gadgets, such as an autonomous leaf vacuum/shredder, and a few robots, there are mostly posters—lots and lots of posters.

"That's a trend we've been seeing," says Gail Dundas of Intel. "There are not as many 'things.' In parallel to the mainstream movement away from hands-on tinkering, ISEF is becoming, year by year, a more cerebral experience, with lots of posters describing research and fewer prototypes that go pop *zip-zap-pow.*"

But what stories some of the posters tell.

There's a kid who trained honeybees to take the place of bomb-sniffing dogs. One student developed a program to help blind people navigate using iPhones; another made a football helmet that detects concussion-level impacts; a few others are getting energy from mud or paint.

And then there's Ari Dyckovsky, who managed to untangle the esoteric phenomenon of quantum entanglement. Dyckovsky developed a theory for an entanglement configuration that could potentially be used to send information (in the form of quantum bits, or qubits) via quantum teleportation. Using quantum teleportation, a spy agency could send encrypted information without running the risk of interception because no particles would actually travel to the new location. The information from one atom would simply appear in another atom in the new location when the quantum state of the first atom was measured.

When the room opens to the public, the kids go through their pitches with any and all passersby, whether a judge, a journalist, a competitor's dad, or just an interested schlub. I stop and talk to Jan Matas, part of a team of three boys from Slovakia who have built—yep, actually built—a victim-seeking robot for use in earthquakes and other disasters. It scans for higher-than-ambient temperatures to detect living things and has a microphone "like a droid. But they are under our control," says Matas, who adds that the team tested it in a former elementary school. "But we found strange individuals living there; it wasn't safe. So we took a baseball bat just in case. In sixteen out of twenty cases, we found the victim."

Some of the projects are dull; some are cool but not particularly practical; some are jaw-dropping, world-changing discoveries. Andy Arthur, an oncologist who has been judging at ISEF for thirty years, brings me to the display of Michigander Nithin Tumma. Tumma discovered that a single protein in breast cancer cells plays a decisive role in turning these cells from bad to worse. When that happens, the diseased cells escape the body's anticancer protections and spread to other organs.

That in itself is an extremely significant discovery. But Tumma didn't stop there. He got the cancer cells growing in petri dishes, then he figured out a way to shut the rogue protein down and watched the cells switch from malevolent mode to a more docile state that was less likely to spread.

"This is by far the most eloquent, impressive thing I've seen in years," Arthur told me. "I have absolutely no doubt that in ten years, this is how we'll be treating cancer."

I reach the physics section as a crisis is unfolding at Taylor's booth. Ochs is trying to calm Taylor, who, in the hubbub of his Hawaii trip, forgot to ask Kenneth to bring along his prop, the chamber that slips into the dense plasma focus (DPF) machine and secures the medical-isotope-generating device. Taylor couldn't drag an entire DPF accelerator to the fair, so his project mostly consists of a poster detailing the concept, the methodology of the proof-of-concept tests, and the results.

"The judges don't need to see that prop, Taylor," Ochs says. "They're physicists; they can understand what's important, which is that you've been able to make medical isotopes by hitting them with five-million-volt ion beams using dense plasma focus instead of a cyclotron or a linear accelerator. *That's* what's important!"

Taylor isn't having it.

"No!" he says. "The chamber, that's where the magic happens. Yeah, they can read the poster, and they can understand it, but they need to *hold it.*"

Ochs says he'll see what he can do. He heads for the door to track down Kenneth. Ochs is right, of course; the physicists judging Taylor's project would certainly understand it, on an intellectual level, by reading the poster. The prop that was left behind in Reno is nothing more than a machined and welded assembly.

But Taylor is right too; like all great showmen, he understands the transcendent value of having a physical object to give to the judges, something they can hold and turn over in their hands, something that triggers a physical and emotional connection—something that makes the idea *real.*

"The big question for the judges is, how much of the work is the kid's and how much is someone else's? To figure that out, we ask them a lot of questions."

The next day, I'm walking through the hall with Chuck Yu, chair of the twelve hundred judges who are assigned categories depending on their subject expertise. Yu, who sports a beard, glasses, and a tie that reads $E = mc^2$, believes that the science-fair system has helped to offset the general decline in educational resources for the sciences. He says that about 20 percent of the entrants are published or hold patents, and the projects are increasingly sophisticated. "By the time they are in high school, they're solving problems that have puzzled scientists for years and doing things where twenty years ago people were doing them in graduate school."

We observe from the periphery as the judges make their rounds, stopping to chat with crisply dressed teenagers. Some contestants are deep in presentation mode with the judges; others chat with neighboring students; still others practice their presentations by talking into thin air. "A lot of them are nerdy," Yu says, "but more and more are outgoing, erudite, and dynamic. More and more, kids who do the science fairs are cool. The ones who do best are the ones who have gotten a lot of encouragement from parents and teachers. And mentors, especially mentors. Working with a good mentor is really fundamental to success."

We approach the physics section, where kids near Taylor stand alone at their booths with their hands clasped, waiting, as Taylor, surrounded by half a dozen judges, explains his invention. Though the all-important chamber, his prop, has not yet arrived (Tiffany sent it the previous afternoon via Federal Express), Taylor works the crowd of judges skillfully, with a confident, if slightly arrogant, charm.

"When he started, he was the only one dealing with fusion," Yu says. "Now there are four or five and they all know him and used him as a mentor. It's hard to overstate what an inspiration he's been; he's really popularizing science."

I tell Yu that the judges seem just as eager to meet Taylor as the kids are.

"If not *more* interested," Yu says. "The Nobel Prize winners can walk around unmolested, but not Taylor.

"But . . ." Yu says, then pauses for a moment. "I hear that he might not attend college. I think college is important. I think the beer drinking and socializing and chasing girls are as important to development as is the knowledge."

The judges separate for a minute to allow a video crew to move in for a quick interview with Taylor. Yu and I step a little closer to listen to Taylor's sound bite.

"Some people may not go into science because they think just nerds go into science, or science isn't cool," Taylor tells the interviewer. "But the thing is, science is cool, and me and my friends who do science are cooler than the people who don't. So if you really want to change the world, go into science. Because that's the future and that's what will really change the world."

It's not particularly eloquent but it does the trick. Taylor, smiling apologetically, turns back to the judges.

Of all the teaching and mentoring and coaching Ochs has done, nothing excites him more than ISEF. "I love the personalities of the kids," he says, "and the energy, and seeing the excitement."

At lunchtime, Ochs finds a quiet spot outside for his flock to gather for a debriefing. Sans Taylor, the mood is significantly calmer.

"Any comments from the judges that stand out?" Ochs asks. "What do they seem most interested in? Do they seem skeptical of anything?"

One of the students, Cole, is a rancher's son whose invention regulates the flow of the bubblers that fill cattle troughs. He's down-to-earth, humble, and grateful. He says he knows that his documentation is marginal and that he's not going to come close to winning. But, he says, "I'm just happy for a ticket."

Claudia, from Lovelock, Nevada, is a bit more ambitious. "I just want to get my name called once," she says.

(It won't happen this year, but next year she will make history outside of ISEF by being the first person from her hometown to attend Stanford.)

When Taylor arrives, the energy shifts.

"It's not here yet!" he blurts out, his face pinched. "Will it get here on time? The final judging is this afternoon!"

Ochs looks at the others and doesn't say the obvious — *We're dealing with a bit of a drama queen here* — then motions for Taylor to sit down next to him.

"Taylor," he says. "Stop. Look at me."

He holds up two fingers.

"First, deep breaths. Next, calm down. It's okay, one way or the other. The main thing is: focus. You've done everything you can. Now let it happen. When you go back in there, focus on what you need to do *now.*

When the judges approach, you do an inner focus to calm your body. Then you listen, and pause. If you start talking right away, you haven't focused yourself on what you're about to say."

When Taylor was eleven, he had a future vision of himself as a nuclear physicist making medical isotopes. Now, seven years later, the vision is doubling back on him. This is Taylor, the eleven-year-old again, trying to save his grandma's life.

Taylor begins to calm. Then, he nods and heads back inside.

"I've never seen him this nervous," Ochs says. "But I understand it. His grandmother's cancer and her death really impacted him. This is the big one for him. It's what he wanted to do all along."

Taylor started out four years ago winning fourth place. Then third. Then Best in Category, along with the Young Scientist Award, which recognized his science project as one of the best in the world.

"And now he wants that top award, the Gordon Moore Award," Ochs says. "This is his big moment, the thing he's been going for. He doesn't want to fail."

Kenneth arrives with the just-delivered chamber and with news. Tiffany and Joey, who had planned to fly in for the awards ceremony, have decided not to come. "[Taylor will] be disappointed," Kenneth says. "It could be his big moment. But . . . will he win it all? We don't know. It's a long way for them to come."

Once Taylor gets his prop, he feels more confident. As the judging enters its final phase, only entrants and judges are allowed into the hall. And so word of the next crisis arrives via messenger: the girl at the booth next to Taylor's runs out and informs us that Taylor has forgotten to eat and he's losing energy, feeling faint. She's sent back in to tell him that food is on its way. People are mobilized; food is found; a delivery method is arranged. Ochs accosts a judge on his way in and asks him to deliver it. But when the judge sees who Taylor is talking with, he leaves the food with Taylor's neighbor in the next booth and returns to Ochs.

"My God," he says. "He's in there with Dudley Herschbach."

Herschbach, the Nobel Prize–winning professor, is trying to talk Taylor into coming to Harvard instead of taking the Thiel Fellowship.

"Taylor," the professor says, turning Taylor's isotope chamber over in his hands, "if you skip college, you'll be missing out on things like parties and girls and football games. And expanding your world and discovering and getting excited about things in whole other fields."

Taylor doesn't tell Herschbach that he already made his decision, during the trip to Hawaii. Taylor isn't going to Harvard, or any other college — at least not now. Instead, he will accept the Thiel Fellowship so he can focus on developing his national security and medical-isotopes inventions. He's betting, mostly, on freedom; he doesn't want to deal with someone else's agenda, has never wanted to. He's dropping out of the education system, like Joey. He has the Thiel Fellowship in his pocket; he's been to Hawaii and hobnobbed with the venture capitalists, seen the possibilities. It's hard to imagine what more college could give him.

The next day, everyone files into the packed hall, set up with thousands of chairs facing the high stage. Taylor takes his place, joining the rest of the contestants who are wondering if their dreams of science-fair stardom will come true.

Again, the Nevada team members sit together, waiting, as the emcee goes through all the categories. And then . . . "The first award in the Physics and Astronomy category goes to Taylor Ramon Wilson, of Reno, Nevada, USA."

Taylor trots up to the stage to accept that award and an additional award from the American Association of Physics Teachers and the American Physical Society. Then he sits back down. The announcement Taylor has been waiting for comes at the very end of the awards ceremony.

Again, it's Wendy Hawkins, from the Intel Foundation: "The year's top ISEF winner, the recipient of the Gordon E. Moore Award is . . .

"Jack Andraka, from Crownsville, Maryland!"

The fifteen-year-old high-school freshman developed a dipstick-sensor test for early detection of pancreatic cancer. His method, which costs three cents and takes five minutes to run, is dozens of times faster, much less expensive, and a hundred times more sensitive than current tests.

Taylor, Kenneth, Ochs, and the other contestants leave the auditorium and drift back to the booths to pack up their exhibits. Unlike it was during setup, when excitement and anticipation were high, the hall is calm; the contestants, whether they won or not, seem grateful to finally be able to relax.

Taylor lets out a sigh as he folds his tri-panel poster.

"I got beaten out," he says, "by a freshman."

29

SCOTCH TAPE

TAYLOR AND KENNETH left quickly after the awards ceremony and flew to Florida, where they'd been invited by visionary entrepreneur Elon Musk and NASA to watch the launch of the first SpaceX Falcon 9 rocket that would deliver a payload to the International Space Station. They woke and had breakfast at 3:00 a.m., then joined Charles Bolden, the head of NASA, on the bleachers to watch the launch.

Unfortunately, the liftoff was scrubbed during the final countdown, just as it had been during the Wilson family's visit to Cape Canaveral a decade earlier. Taylor tried to convince his dad to wait for the next day's launch window and skip the Davidson graduation, but Kenneth wouldn't have it. Davidson, he told Taylor, had been so much a part of his life, and had given him so much.

But as they streaked their way west, Taylor's mood darkened. Though his medical-isotopes project had won the top physics award and two other awards, the recognition he'd dreamed of for the project that meant the most to him had slipped out of his grasp.

"I lost," Taylor told everyone when he got back to Reno.

That made no rational sense, of course. Taylor had won first place in the Physics and Astronomy category. He had, over the past four years, walked away with nine awards, including two first-place awards in the physics category and the Intel Foundation Young Scientist Award. He'd won more than a hundred thousand dollars in prizes, and a trip to Switzerland. He had patents, marketable ideas, and people lined up to invest in them.

Kenneth tried to put things in perspective for his son, then gave up, figuring that once Taylor got to his graduation and started celebrating with his friends, he'd snap out of it.

But that didn't happen.

"I haven't seen him this bummed out since Fukushima," math teacher Darren Ripley says. "But after the meltdown, he snapped out of it and started getting optimistic when he saw the opportunity for making reactors safer with his modular design. But when he got back from ISEF, he was crestfallen, and it didn't go away. He felt like he'd failed."

Ikya, too, was facing a major failure for the first time. She'd been denied admission to Wharton, her first choice.

"And yes," she said, "I'm taking it personally."

She'd be going to Berkeley instead. Sofia would be going to NYU. Even Ron Phaneuf would soon be leaving, heading off to retire in the Colorado Rockies. And Bill Brinsmead, though he had no plans to leave town, was phasing out his involvement at UNR.

Over the summer, Brinsmead and Taylor would inspect a few commercial properties in Reno. Apart from that, though, Taylor's business plans seemed to be stuck. When Thiel administrators and others would ask how he was doing, Taylor would offer vague, unconvincing answers. "I'm looking at a first round of financing soon," he'd say; or, "I'm considering a management team."

Plenty has been written lately of the pro-youth, anti-age bias of Silicon Valley. But the flip side of that bias is the questionable expectation that genius tech kids are automatically genius businesspeople too. Thus far, Taylor had been able to leverage his boy-genius image very successfully. But now, he seemed trapped by it—trapped by his need to know the answers rather than ask the questions; trapped by his reluctance to make the (completely reasonable) admission that, as an inventor and scientist, he didn't know the first thing about starting a business. For the first time since I'd known him, maybe the first time in his life, he seemed unwilling to jump in and just do something that might fail.

Meanwhile, the big dreams of fusion energy were smashing into hard realities. That fall, the National Ignition Facility (NIF) announced that it would not meet its goal of igniting a fusion plasma any time soon. Under budgetary and political pressures, NIF began shifting its research focus toward nuclear weapons instead of fusion energy.

The U.S. contribution to the ITER fusion research project was uncertain too, as Congress threatened to hold up funding just as construction was ramping up. Suddenly, the pessimistic joke about nuclear fusion being "the energy source of the future, and it always will be" seemed like more than just a naysayer's cliché.

A few weeks after graduation, Taylor texts me: *Big news. Call me!*

"A couple of years ago I had this kind of crazy idea," he says when

I reach him. "And I finally got the parts together and did the experiments. I've been in the lab for the past two weeks solid and I finally proved it: I made Scotch tape produce nuclear fusion."

I wonder if I'm hearing him right.

"This is a *big* discovery, a *Nature*-type story," he continues. "What's going on is that when you pull tape off the roller you're breaking the intermolecular bond and generating very high voltages with a very small charge." He tells me that he's built a mechanism to pull the tape at a constant rate under vacuum with some deuterium in the chamber. By tweaking the variables, he's been able to generate a hundred thousand volts, enough to accelerate the ions into fusion.

"I'm getting constant neutron output on run after run, way over background levels, and I can see the x-rays coming out. This could represent the cheapest source of neutrons by far."

Now I'm really wondering if he's lost his way. Scotch tape experiments? What about nuclear security? Medical isotopes? Fusion energy? Next-generation fission? What about building his company?

"This is just the funnest part about science," he says. "Where you hypothesize something and you plan for days and get a whole experiment set up, and you see your data and find out that your intuition was right, that it held up."

He adds that he's been invited back to TED, which is coming up in a few weeks, and that he's going to make a "big announcement" there. I try to get more out of him, but he wants to keep it a surprise. "This will be the first TED talk I ever planned," he says, "and it's gonna be really big!"

I'm just hoping it won't be about Scotch tape.

Tiffany and Kenneth can see their son drifting too. "I'm not so sure the fellowship was a good idea, skipping college," says Tiffany. "He's constantly researching, but not really doing anything. I'm wondering, *Do you really want to start a company? There's so many things you can do right now.* I think Taylor likes being a kid, and is having a little trouble with figuring out what he wants to be, where he wants to be."

Who, in his situation, wouldn't be? Once you've achieved everything he did at seventeen, where do you go from there?

After Pittsburgh, I don't see Taylor for several months. In mid-December, we meet in Los Angeles and start with a tour of the SpaceX factory, where Elon Musk and his team are building the next generation of orbital rockets. Taylor, who knows Musk from the past two TED conferences (in 2014, they showed up dressed almost identically in blue gingham button-down shirts), has always been impressed with him, although, Taylor says, "I think we have sort of clashing personalities. But you can't help but be impressed. He's the only person who's taken Silicon Valley capital and built a massive industrial complex. Look at the risks he's taken. I like that."

But it's Taylor, not Musk, who is recognized the moment we walk into the lobby, by a man who's part of a crowd of NASA engineers and one astronaut—distinguished by his blue NASA jumpsuit—who are here for a training session.

"Hey, you're that kid who . . ." the guy says.

The astronaut pipes up: "Oh yeah, the fusion kid—you were on CNN! What you're doing is really cool. I've always been interested in nuclear fusion, especially for rocket propulsion."

Taylor grins broadly. "And I always thought I'd be an astronaut," he says.

Now eighteen, Taylor has been traveling by himself more often. That evening I'm driving him to Burbank when the song "Radioactive," by Imagine Dragons, comes on the radio. To me, it's the worst sort of over-affected power pop, but Taylor loves it. He turns it up and sings along at the top of his voice. I ask him if it's his favorite song and he grins broadly. "What do you think!" he yells. "Of course it is!"

When the song's over, he tells me, "You know, you really don't have to do all the driving." Taylor finally got his driver's license when he started doing the proof-of-concept tests of his medical-isotopes device at UNR.

"My dad put his foot down," he says. "He said he wasn't going to drive me all the way out there and back every day."

We've been invited to that evening's live taping of *The Big Bang Theory*, the TV situation comedy that follows the idiosyncratic and narcissistic physicist Sheldon Cooper and his relationships with his small circle of equally nerdy scientist friends and with Penny, their cute, nonscientist neighbor. "You know, I think there actually *is* a part of me that's a bit like Sheldon Cooper," Taylor says as we approach the Warner Bros. studio. "There's the not-driving thing—or at least there was. And I'm also a bit of a hypochondriac. Oh, and I can be kind of stubborn."

The show's popularity has thrown a questioning light on the

premise that science can't be cool as well as the contention that Hollywood can't be accurate when it comes to science. Director Mark Cendrowski has invited us, but we spend much of the preshow time talking with the show's science adviser, UCLA physicist David Saltzberg. Saltzberg checks scripts and supplies dialogue and mathematics equations. He also checks one of the show's central props, an ever-changing whiteboard covered with scribbled equations and diagrams —"to make sure it's not a bunch of faux-scientific gobbledygook," he says.

"By the way, we've got a show coming up that uses the word *tokamak,*" he tells Taylor. "So if you want to try your hand at putting something on the whiteboard, let me know."

Saltzberg and Taylor talk shop about each other's projects until the bombastic emcee starts to get the crowd riled up by inviting people to come up and dance. There's a Hasidic guy from Brooklyn, a large Hawaiian guy who jiggles as he dances, and a young girl wearing a shirt that reads *I ♥ Coitus,* a riff on one of the show's ongoing themes, the sexual awkwardness of science geeks.

Sitting through a live studio television taping can be an excruciating experience. The emcee reminds us repeatedly to laugh "naturally" on the third or fourth takes of scenes, "just like you're seeing it the first time!" Taylor calls me out for not keeping up the charade of artificial enthusiasm, reminding me that "these guys are doing great work promoting science — it's worth supporting!"

Between scenes, audience members are invited up to say why they love the show. One woman introduces herself as a high-school science teacher. "The show gets my kids interested in science," she says. "Jim Parsons [who plays Sheldon Cooper], he's a geek, he's a nerd, he is so cool!" A teenage girl, playing up her science-nerd look, says the show "finally made it cool to be a geek and nerd. This has made me so happy!"

When it's over, as the studio empties out, Cendrowski and Saltzberg invite us onto the set for the cast and crew's holiday party. The living-room set is packed with the show's familiar astro-nerd paraphernalia: old scientific instruments, bookshelves full of science tomes, and an educational beach ball developed by NASA printed with a sky map of temperature fluctuations of the radiation left over from the big bang. Cendrowski encourages Taylor to sit in Sheldon's favorite seat, and

Taylor eases onto the couch and takes a satisfied look around. Suddenly, his eyes zoom in on a prop on the shelf. It's a Korean War–era Geiger counter that's been there since the show's debut in 2008.

"Y'know, they were supposed to remove the radioactive check sources when they decommissioned them, but lots of times they missed them," Taylor says, rising from the couch and walking toward the shelf. "Mind if I pull it down?"

Cendrowski raises his eyebrows. "So it could be still . . ."

"Radioactive?" Taylor says. "Yes, there's a good chance. Want me to check it out?"

The set's leadman (who supervises set-dressing) runs to fetch a screwdriver, and Taylor begins to dismantle the device. Alerted to the drama, the show's co-creator and executive producer Chuck Lorre grabs cast member Johnny Galecki; they arrive just as Taylor pulls off the outer cover. Then he removes an internal panel and exposes a small white disk.

"Bingo!" he says. "Strontium-90."

Lorre's jaw drops, as does Cendrowski's.

"Oh. My. God," says Saltzberg.

"Maybe," says Cendrowski, "it would be a good idea to . . ."

"Get it off the set?" Lorre says. "Like, really quick?"

"Tell you what," Taylor says, looking up. "How about if I pull it out of there and take it home? Then you've got a safer environment, and I've got myself something new for my collection."

Lorre nods, while Cendrowski and Galecki shake their heads in amazement.

"Taylor," Saltzberg says, laughing, "I think you've just become a permanent part of this show's lore."

A few months later, Taylor and I rendezvous in San Diego for the International Atomic Energy Agency's twenty-fourth Fusion Energy Conference. The FEC brings together researchers from around the world every two years to discuss developments in fusion experiments, theory, and technology. Although much of the program is related to some aspect of ITER and its magnetic confinement approach, the conference also attracts researchers working in inertial confinement fusion, hybrid systems, and several wildcard methods.

Taylor, who arrived a day before me, is extremely fired up when I catch up to him. "I'm not a theory guy," he tells me, "but I am at a theory conference, and I'm loving it!"

Taylor has spent the past day rubbing elbows with the PhDs who work in the billion-dollar labs that spawned the breakthroughs Taylor read about when he was a child, before David Hahn's errant exploits spurred him to tackle hands-on nuclear physics himself. As we stroll around, it quickly becomes obvious that Taylor has now become an esteemed, if honorary, member of this club of elite physicists. Everywhere he walks, researchers recognize him and call out to him. Some mention his TED talk; others congratulate him on his more recent talk at the National Ignition Facility; still others talk about his television appearances and the renewed interest in nuclear fusion that he has sparked.

The subject of Taylor's most recent TED talk—"Reinventing Nuclear Fission"—surprised me, and many in the audience. He was taking a risk promoting a new fission technology so soon after the Fukushima disaster, when much of the world was running away from fission power. But Taylor told the audience that even though fusion will transform the energy industry in the long run, the world's shorter-term energy needs could be met by a new, safer generation of fission reactors. He presented his vision for small, modular fission reactors that could be built in factories rather than onsite and shipped by train or truck to their final destinations. These would be molten salt reactors with no weapons-grade material inside and no way for the reactor to melt down or for radioactive material to leave the vessel. They'd be installed underground and would run for about twenty years without refueling before shutting down and self-sealing—thus solving the problems of waste processing and transportation and minimizing safety risks.

"The perfection of fusion and fission in the twenty-first century, that's what I'm aiming for," Taylor said.

This talk was the first presentation in his life that Taylor had actually prepared and practiced for, and as a result, it seemed a little stilted; it didn't come off as smoothly as when he was winging it. "But in general," Taylor says, "TED was so much fun this year. I knew what was going on and I got to see people I knew, and it gave me the opportunity to take the modular reactor project public." After the talk, Taylor was approached by several potential investors.

His speech at NIF was more casual and more fusion-focused. He told the charmed crowd at the largest experimental science facility in the U.S. about how he had started to build his reactor at eleven and finally succeeded at fourteen. Then he moved into his nuclear medicine and antiterrorism projects. "I know you all have some experience in that field!" he joked. He talked about his collection of radioactive materials and his desert quests for uranium ore and pieces of nuclear history. Then he turned to energy and to the philosophical nature of science. "It's human nature to pursue the big, challenging things in the universe, like space and fusion," he said.

The talk was a big hit, judging by the number of congratulatory remarks Taylor receives as we board a bus for a tour of the General Atomics DIII-D tokamak research facility, in the hills north of San Diego. Although Taylor has been there once before, he's convinced that I need to experience it. "It's incredibly cool; you can't leave San Diego without going there," he says.

Taylor tells me about last night's update on ITER as we walk from the bus to the General Atomics facility. "I'm totally brimming with ideas — it seems like something sticks out in every talk, something that I need to do an experiment for."

A minute after we're waved in, a General Atomics official named Brenda Bowman comes running up, obviously irked with us. "You didn't sign in!" she yells.

We'd been told that U.S. citizens didn't need to show ID or sign in.

"Who told you that?" Bowman demands.

Not wanting to get anyone in trouble, we try to shrug it off, and we head back toward the sign-in table.

"Wait!" she says, reaching for her walkie-talkie. "This is a major security violation."

Just then, another GA official appears and touches her arm lightly. As he whispers something into her ear, a look of horror comes over Bowman's face, and she covers her mouth.

"I am *so sorry*," she says to Taylor. "I didn't realize who you were at first, but now I recognize you. I really, really apologize. I'm a big fan of yours and I appreciate all you're doing for fusion, and for science."

Taylor handles it graciously, tells her it's no problem, and apologizes for misunderstanding the entry instructions. "Really, it's no big deal," he says. "You're in charge of making sure people are identified and accounted for at a very important facility. Why should you treat me differently than anyone else?"

I realize that I am seeing a very different Taylor. This Taylor seems much more mature, gracious, and—I never thought I'd use this word to describe him—humble. I can't help but wonder if "losing" at ISEF (though I would argue that he didn't lose) was the best thing that could have happened to him.

Taylor's humility is all the more surprising considering that he's playing in a bigger arena now. At one point, I approach as he's talking with an important-looking man in a suit; Taylor waves me over and introduces me to the head of the International Atomic Energy Agency. Later, he introduces me to Ed Moses, the head of NIF.

That afternoon I see him surrounded by half a dozen physicists hanging on his every word. "We don't have anything to inspire kids," Taylor tells them. "We don't have an Apollo program, we don't have something new like computers when they first came out. Going to the moon inspired a generation to get into aerospace. People like Bill Gates and Steve Jobs got a generation of kids to get into computers. There's this historical precedent; people are looking for something to be inspired about."

He continues: "Did you see that proposal Obama made at Argonne [National Laboratory] about sending a percentage of money from oil and gas leases to develop advanced vehicles? It was only two billion dollars over ten years! I watched that and I thought, *Is this the best we've got?* Why not take *all* that oil and gas revenue and do something really amazing with it? Why not do what Kennedy did and say something like 'By the end of the decade, we will have nuclear fusion'?"

As I watch the lit-up faces all around him, I can see that what is dawning on me is dawning on everyone there.

Among the scientists gathered around Taylor is IAEA fusion physicist Ralf Kaiser. Taylor introduces us and says, "You guys should have lunch." Kaiser seems keen to talk with me, but he's booked for lunch; we make plans for breakfast the following morning.

I'm thinking Kaiser wants to sell me on the viability of fusion energy, as is often the case when a fusion scientist catches a journalist by the ear. Instead, the next day, I find that Kaiser wants to talk about Taylor.

"As a student in Germany," he says, "I was in a special program for profoundly gifted students. You see, I *was* Taylor at one time. What strikes me about him is that there are only so many in each generation. These are the people who are more likely to make discoveries and affect society disproportionately. Taylor wants to create, and he wants to achieve something for the world, something beyond himself. Thus, the most important question is, which direction will he choose to go?

"Fusion is hard to do," Kaiser continues, "but the technical problems are not as likely to kill it as its image problem. Why does it have such an image problem? Because of a lack of education. I can't tell you how many times I've stood in a pub with a pint of beer and talked with someone and told him what I do. It starts with, 'Isn't that dangerous?' After a half hour, it's 'Why exactly are we not doing everything we can do right now to make that happen?' "

He pauses. "Did you notice that the Nobel Prize in Physics was announced two weeks ago?" I tell him that I did, and he asks if I remember the co-winners' names. I'm embarrassed; I don't.

"I also don't," he says. "I can tell you exactly what the scientists did, all about the potential applications of their research. But," he says, "I can't tell you their names! I don't remember! Considering who I am and what I do, that's pathetic! Unfortunately, it's the way the world is. But I hope it's not the way it is going. Because which society will be more successful? The one that encourages scientists, or the one that encourages stockbrokers and movie stars?"

I ask him how society can best encourage young people to become scientists.

"We will have to make them famous, turn them into stars," he says. "And this is where Taylor comes in. People always ask, why isn't science cool? Well, with Taylor, it *is* cool. He is really an inspiration for young people—but not just young people! Everybody is inspired by him, and already more people are interested in physics because of him. Maybe someone like Taylor is just what we need right now. Maybe this will be where he really shines."

On my last trip to Reno I pay my visits with a particular question in mind: What if Taylor had never come to Davidson Academy? What would have happened to him if he'd stayed in the public schools in Texarkana?

"He'd either fall through cracks and feel disillusioned and disenfranchised by the traditional education system," says math teacher Darren Ripley, "or, well, there are some very bright mass murderers out there, your Ted Kaczynskis. Just joking—but when Taylor makes his first hundred million, I want a cut!"

Bob Davidson is more serious. "He would have inevitably ended up as a high-achieving, successful person, mostly because of his parents. Much like our philosophy, they never limited him. But it wouldn't have been as easy or early, and with as much understanding."

Taylor had achieved more than would seem possible in his first eighteen years. And yet, his story began much like David Hahn's, with a brilliant, high-flying child hatching a crazy plan to build a nuclear reactor. Why did one journey end with hazmat teams and an eventual arrest while the other continues to produce prizes, patents, television appearances, and offers from around the globe?

The answer is, mostly, support. Hahn, determined to achieve something extraordinary but discouraged by the adults in his life, pressed on without guidance or oversight—with nearly catastrophic results. Taylor, just as determined but socially gifted, managed to gather into his orbit people who could help him achieve his dreams: the physics professor; the older nuclear prodigy; the eccentric technician; the entrepreneur couple who, instead of retiring, founded a school to nurture genius kids who might otherwise be overlooked. There were several more, but none so significant as Tiffany and Kenneth, who overcame their reflexive—and undeniably sensible—inclinations to keep their son on the ground. Instead, they gave him the wings he sought and encouraged him to fly up and reach for the sun.

I drive over to the Hub café and park behind the electric hearse out front. "Regional science fair's coming up," Brinsmead tells me, after I grab my coffee and sit down, "and Taylor's borrowing the van so he can bring some stuff over."

Brinsmead smiles. "Yep, he's starting to learn about mentoring. Last week, he brings in this wide-eyed Indian kid who wants to make a cloud chamber for ISEF. And there's also a Thai kid I'm helping; he was on Fusor.net and Taylor introduced us. He's sort of like Taylor Version Two, but in Thailand. Last time I went over there I brought a whole bunch of parts for his fusor, and Taylor coughed up a neutron detector for him."

Brinsmead asks if I've heard the latest from Tiffany and Kenneth. I tell him that the Wilsons' house is my next stop.

Bill slips his spoon into his coffee and looks down, silent for a moment.

"When he got here he was such a little guy," he says, shaking his head. "And to think it all started when he was even littler, stirring up his grandma's pee and getting that crazy idea . . ."

I'm in the car when Taylor calls, as excited as I've ever heard him. "Did you hear about NIF?" he says. "They've done it! They hit scientific break-even with one of their shots!"

He reads the news report to me: " 'U.S. scientists announced an important milestone in the costly, decades-old quest to develop fusion energy, which, if harnessed successfully, promises a nearly inexhaustible energy source for future generations. The National Ignition Facility, in Livermore, California, has generated the first break-even nuclear fusion reaction.'

"I should call some of my friends there!" he says.

It's obvious that Taylor won't be calming down for a while, so Tiffany and Kenneth and I go out for a meal. They've made a big decision, they tell me: They're moving back to Texarkana so that Joey can be with his friends and, hopefully, find his feet again.

"We had some meetings at the high school there," Tiffany says, "and they were honest that they've got nothing to offer him academically. So he's going to finish up through the Nevada homeschooling system. But he'll be able to go to school activities in Texarkana and keep busy socially."

"He's also opening up a little more about college," Kenneth says. "Apparently, several people told him he'd be good at engineering and would enjoy it. Now he wants to look at a couple of different universities."

I ask what Taylor will do. Kenneth says they're not sure if he'll go with them or stay here; he's still deciding. But he tells me that Taylor's been offered a grant from the Arkansas Development Corporation if he locates his company there. Kenneth and Taylor recently flew to Little Rock to meet with the governor, the head of the University of Arkansas, and the leadership of Stephens Inc., the nation's largest off–Wall Street investment banking firm, based in Little Rock.

"They're the ones who got Walmart and Tyson going," Kenneth says, "and they offered to help Taylor get his business up and running. They really took a liking to Taylor, and we had a lot of laughs." At one point, Curt Bradbury, who heads Stephens, said, "Here's the governor and these four men in their sixties, at the pinnacle of their careers, and we're all at the mercy of the eighteen-year-old boy in the room."

"I'll tell you something," Kenneth says. "Taylor's given us a lot of challenging moments, but a lot of fulfilling ones too."

I find Ron Phaneuf in his laboratory, slowly packing up. "It's getting a little lonely around here without him," he says, glancing toward the fusor, which still occupies its spot near the corner, surrounded by yellow radiation-warning signs and tape.

"It was very fortunate happenstance that brought Taylor to us," Phaneuf says. "He became such a catalyst for so many things around here."

I mention to Phaneuf that, at his graduation ceremony in the spring, Taylor told Jan and Bob Davidson: "Of all the things you've done for me, the one I'm most thankful for was that you gave me the opportunity to meet Bill and Ron."

Phaneuf smiles and shakes his head. "I can tell you one thing, and I know Bill will tell you the same: We learned as much or more from him than he learned from us."

I ask him to tell me more about what he learned.

"The strength of ideas," he says. "Like he said in that TED talk, 'I just believe I can do it; I don't accept that I can't do it.' That's his philosophy of life. He's got passion and a sharing personality. And all those ideas. And now he's turning some of them into things you can sell. But mostly, he's selling science."

Phaneuf's phone rings; it's a quick question from one of his two last postdoctoral fellows. When he hangs up, he motions for me to follow him over to Taylor's cluttered side of the lab. "Have you seen this?" he asks, nodding toward a contraption hooked up to the fusor. It's got neutron detectors and other instruments and a vacuum chamber with a motor inside that's attached to . . .

. . . a roll of Scotch tape.

"This has to be the most heavily instrumented roll of Scotch tape in history," Phaneuf says, laughing. "But you know what? He really discovered something."

So it wasn't so crazy after all?

"Not at all. When he started that I thought, *What's he doing?* But it's related to the optical phenomenon of triboluminescence, which generates light when you stress materials and break chemical bonds. It's not unknown, but it's not well understood—and this is the first time anyone proved there were energetic particles coming off it.

Taylor used a magnet and observed the bending. Those results are pretty convincing.

"One thing I've learned is, don't underestimate Taylor."

Now it's Taylor who drives, toward the Reno airport. He's got a lot of speaking engagements coming up, and he's enjoying traveling and meeting people and "getting people inspired about science, which is really the point of all of it."

I ask him if he's got a regular, prepared talk that he typically gives.

"No, I still don't prepare anything," he says, laughing. "It's always different. The world is always changing, and so is my talk. Science is always making new discoveries. Just look at NIF; that'll probably be in my next talk."

I ask him something else that's been on my mind: Now that he's an adult, has his approach to science changed?

"I think so," he says. "At first I was exploring things because I was interested in them. It was fun, but in a way it was selfish too. Then there was the thing with Grandma dying, and it turned out that I was really good at science, and it became a responsibility. I realized I had the capability of doing things that could really change the world. So now I'm doing things for bigger reasons. It's still fun, but there's that responsibility on top of it."

Taylor's early successes defined him as a child prodigy, a boy genius. What, I wondered out loud, would his next act be?

"I always tried to shunt off those labels; they kind of bothered me," he says. "But now I see that kids sometimes have an advantage when it comes to invention, because they have a less constricted view of the world. Older scientists sometimes get this mentality that it can't be done or you shouldn't even try it, whereas kids are not so closed-minded. They can see things in ways that adult scientists often can't.

"I want to grow a business that allows me to create really useful things. But hopefully I'll never have to grow up too much," he says. "Because what makes really good scientists is a healthy disregard for limits and conventions that say you can't do this or that.

"I hope I never lose that."

I say goodbye to Taylor and check in, then move through security, stopping and raising my hands in the scanner (which I now know is creating an image from radiation emitted from my own human body).

My head swirls with memories: checking the box of radioactive bomb parts onto the plane, making yellowcake, feeling the heat of the Father of All Bombs in my face . . . and visiting the place, if not the moment, where a small boy stirred his dying grandmother's pee and came to the outrageous conclusion that the world revolving around him was one that he could actually change, and improve.

It's not just the world now, it seems, but a whole solar system revolving around Taylor, this kid who dared to believe he could literally create his own star. I'm in his orbit too now, I suppose, and as far as transformations, I have not been immune. What I learned most profoundly from Taylor's journey is that our most energizing insights can arrive at the moment when our struggles reach their apogee and burst open, illuminating an image of a transformed future – one that's still imagined, but already filled with sparkling immensities.

Taylor built a star, then he became one. And now he's lighting up the world.

EPILOGUE

THERE'S ONLY ONE Taylor Wilson, but his unique experiences and stellar trajectory contain lessons we can apply to all the gifted ones among us. I happen to have two young geniuses of my own. They attend public school in Ann Arbor, Michigan – the town, you may recall, where 99 percent of parents who responded to a school-district survey identified their own children as gifted.

No one needs a PhD in evolutionary psychology to understand why we parents believe our own children deserve exceptional treatment. And the latest science actually supports our intuition that our children are gifted. A growing body of academic research suggests that nearly all children are capable of extraordinary performance in some domain of expertise and that the processes that guide the development of talent are universal; the conditions that allow it to flourish apply across the entire spectrum of intellectual abilities. Parents, the primary creators of a child's environment, are the most important catalysts of intellectual development. While there's no single right way to rear a gifted kid, talent-development experts say there are best practices for nurturing a child's gifts in ways that lead to high achievement and happiness.

Nearly everyone is capable of extraordinary performance, but not in every domain. The challenge is to find the outlet that best fits a person's unique set of interests and characteristics. As a start, give kids lots of exposure to different experiences in their younger years, and pay attention to what they pick up on.

"Take your kids places," say talent-development experts, whose research has convinced them that the development of creativity and innovation depends on exposure to unusual and diverse experiences during formative years. Early novel experiences play an important role in shaping the brain systems that enable effective learning, creativity, self-regulation, and task commitment.

These experiences can become more targeted when children develop interests they want to explore. Don't be afraid to pull your kids out of school to give them an especially rich and deep learning experience, especially when it relates to something they're curious about. Parents often focus on attendance and grades, forgetting that, especially in a child's early years, grades are far less important than actual learning.

As you help your children make contact with as many activities and subjects as possible, stay on the lookout for signs of strong interest and unusual talent, then move quickly to provide opportunities to develop those passions and talents. Research has shown that these sparks of curiosity are critical windows of opportunity, and they are fleeting; if they go unnoticed and uncultivated, they usually fizzle out, often permanently.

It's worth noting that some children who appear untalented are simply late bloomers. "Sometimes," says psychologist Barbara Kerr, "you have to keep the door open a long time." But once a child finds an intellectual passion, the learning process can accelerate rapidly, as he or she becomes inspired to excel. At this point, the key role (and challenge) for parents is to support their children—without being pushy—and connect them with resources that allow them to extend their interests. The odds are that most kids will get into something less harrowing than hands-on nuclear science, but parents who encourage their children to take intellectual risks must be willing to take some emotional risks and give their children the freedom to discover who they want to be. This can be difficult if a child chooses a quirky path (painting or playwriting instead of medicine or engineering) that seems chancy or radically different from a parent's own path.

Some commenters contend that Taylor's accomplishments wouldn't have been possible if his parents weren't relatively well-off. This argument doesn't hold up. Although most of us are constrained to some extent by financial resources, more money does *not* buy good parenting; usually it buys proxy parenting: nannies and babysitters, material goods, and elite schools and programs. Far more effective than money are parents who put time and effort into their children and their children's interests.

When it comes to parenting, the best things really are free. Kenneth and Tiffany built a base of support that was both intellectual and emotional. They developed customized, hands-on opportunities that meaningfully expanded their children's—and often their own—range of experience, and they reached out to their communities to extend their influence. Most parents instinctively try to keep their children away from things that can kill them, but Tiffany and Kenneth found creative and safe ways to enable and support Taylor's potentially hazardous passions. In the Wilson household, there was a culture of "intellectual spoiling" that allowed Taylor and Joey to pursue their interests as far as they cared to take them.

Child psychologists and educators point out the difference between this sort of ultra-supportive parenting and helicopter parenting, in which competitive, overcontrolling parents steer their children toward the parents' choices. It's far better to provide the children with the fuel they need — supplies, mentors, encouragement — and then let them pilot their own helicopters. In other words, let the child lead.

Eventually, school will be an issue. In recent years, it has become fashionable to delay children's — especially boys' — entrance to kindergarten, even if they are already reading and intellectually ready for school. It's well intended, but research has shown that in the long term, it does a bright child no favors.

Gifted children in the early grades usually do fairly well, especially if they have skilled teachers who are willing to explore learning opportunities beyond standard curricula. But our schools' ability to cultivate diverse talents starts to fall apart in the middle grades — as do many of the kids. It's easy for a creative, talented kid to get lost when attention shifts to disruptive students or to those who need additional help to keep pace with the others. If your child spends much of the school day waiting for the rest of the class to catch up, it's time to intervene.

Not every family can pull up stakes and move across the country so a child can attend a school that perfectly matches his or her interests and learning style. For gifted children to get needed resources in their own communities, parents usually have to become resourceful and assertive advocates.

Your school system may be able to arrange grade-skipping and other acceleration, dual enrollment at a higher-level school, a transfer to a magnet school, mentoring, or other options. Often a school administrator's first response to an inquiry is "We can't do that." In some cases, that may be true; in far more cases, what he or she means is "We've never done that before" or "We don't want to make the effort to do it." Be ready with questions: Is the lack of willingness to take action due to a written policy? At what level are such decisions and policies made?

Is the reasoning rational and clearly supported by facts and solid research? Has the intervention you seek been done at other schools in the district?

By all means, get your kid tested. Test results are a solid backup to a parent's arguments for exceptions to the norm. Some parents may be uncomfortable describing their child's needs in terms of "special education," but the fact is that without specialized services, students at both extremes of the ability scale don't learn well in a normal classroom. Testing can also reveal issues holding a gifted child back; some of these conditions may be disabilities—such as dyslexia or attention-deficit/hyperactivity disorder—for which special services are mandated by law.

Any child with strong intellectual abilities should take the ACT or the SAT college entrance exam in the seventh grade instead of waiting until junior or senior year of high school. A high score can open up all sorts of doors, including invitations to university-run summer courses, scholarships, travel opportunities, and high-quality supplemental learning programs. (For an up-to-date list of strategies, programs, resources, and financial-aid options for gifted children, see www.tomclynes.com.)

Finally: Keep in mind that putting a child on a gifted-education track has pitfalls. The gifted student's needs should be considered in the context of the entire family's well-being. Also, intellectual development shouldn't come at the expense of social/emotional development. Once a child is labeled *gifted,* the constant pressure to perform can become an emotional burden, and it can also impede the development of a growth mindset. Some students become worried about protecting their image and don't want to embrace intellectual risk-taking or open themselves up to failures that could challenge that perception.

Labeling and early achievement come with other risks too, as Taylor and his parents learned. It wasn't Taylor or his family who fixed the labels *genius* and *Einstein* on him, but Taylor, Joey, Tiffany, and Kenneth all experienced some of the darker consequences. A pedestal, as it turns out, makes an extremely wobbly foundation.

Despite the advantages implied by the term *gifted,* an exceptionally talented child's journey can be rocky. Half a century ago, intellectually precocious students were seen as strategic resources, and nurturing exceptional talent was a high priority. Educators recognized that these children had special needs, and policymakers realized that meeting those needs was essential to the country's security. Programs for the academically talented burgeoned during the Cold War and began to pay off as Sputnik-generation brainiacs rose to become high-achieving adults. Though the focus on the country's brightest children was driven by fear and worries of inferiority, the byproduct was a surge in innovations that boosted the quality and length of American lives, created tens of millions of jobs, and fueled much of the West's economic growth.

Now that focus on gifted education has moved east, along with many of the dividends it produced. The world's fastest developing economies have picked up the ball we've dropped and are running with the idea that future success depends on developing the talents of the brightest young people. As American school standards continue to decline compared to the rest of the world's, rising countries have placed their bets on high-performing children and created innovative programs to support them.

For the past three decades, American policymakers have wrung their hands over the loss of jobs and competitiveness while at the same time stripping resources from gifted-and-talented programs and research, ignoring the fact that success in a postindustrial economy depends on the intellectual capability of a nation's population. Instead of trying to rebuild the nation's talent pool, we are squandering a crucial natural resource.

Beyond the threat to any one nation's economic competitiveness, there's the cost to the world of failing to cultivate the talents of those who could push the frontiers of knowledge forward. We'll need the breakthroughs of these potential innovators—in domains as varied as nuclear fusion energy, pandemic prevention, and peacemaking—if we are to stand any chance of preserving a habitable planet.

Ironically, the decline of educational programs for the gifted has coincided with an increase in the understanding of how best to educate gifted children. We can now confidently predict which young children are likely to go on to be high performers if they get the right support, and we know how to nurture their talents—often in ways that are not particularly resource-intensive. And yet, we are not making society-wide efforts to do so.

It is the responsibility of public schools to provide appropriate education to all children. It has become clear, though, that our industrial education system is woefully inadequate for children of all abilities and particularly for children at the top end. A few schools have overturned this broken model and developed new, research-based approaches to pedagogy that can serve the entire spectrum of children. Their approach rejects the one-size-fits-all model and replaces it with an individualized and interest-based learning model. Instead of forcing students to fit into the system—which has never worked very well—the schools try to meet each child's unique needs. Their role is to help students recognize their talents and support them as they pursue their chosen interests.

Some argue that this approach would strain the resources of normal school systems—but so do the behavioral problems of students who are underchallenged or struggling to keep up. And the social costs of an undereducated and underproductive populace strain our entire society.

Recently, after several decades in the shadows, gifted children's needs are creeping back into the national consciousness. As government support has dried up, private philanthropists such as Jan and Bob Davidson have stepped in with funding and innovative programming. Slowly, new options are opening up for gifted kids. Some school systems are reviving—or at least considering—grade-skipping and other acceleration options, and university-run summer programs for high-schoolers have expanded, as have online education programs. New talent searches, competitions, and awards programs—especially for STEM-focused kids—are expanding what has become something of a farm system devoted to finding and developing youthful talent. Though many programs are still hobbled by funding shortages and outdated attitudes, some in the gifted-and-talented community are seeing signs of a budding renaissance.

And yet, one has to wonder if it's the right kind of renaissance. The beneficiaries of the latest expansion of gifted-and-talented programs have largely been the children of affluent, well-educated parents, whose preoccupation with high-performing superchildren seems to be driven, at least in part, by parental vanity. A booming talent-development industry has sprung up to boost these children's chances of success, with legions of tutors, test-prep entrepreneurs, and admissions consultants (one of whom promises to return his sixty-thousand-dollar fee if a student isn't accepted to an Ivy League university).

"In the quest to pinpoint and promote exceptional youthful promise," wrote Ann Hulbert in the *New York Times Magazine*, "testers and contests and advocates may have unwittingly introduced early pressure to conform, not to the crowd but to an assiduously monitored, preprofessionalized and future-oriented trajectory." The early pressure to both excel and conform often comes at the emotional expense of the children whose accomplishments provide their parents with bragging rights. Acquiring knowledge for its own sake has gone out of fashion, replaced by a high-pressure talent track onto which promising, prosperous children are pushed. This track is competitive, highly scripted, and pointed directly at the difficult but somewhat mundane goal of Ivy League acceptance followed by careers that promise to be lucrative, if not necessarily rewarding or world-changing.

Maybe that's not what education for the gifted should be all about. Maybe it should be about (1) encouraging supersmart kids to discover who they really are and who they want to become and giving them the support and freedom to pursue their passions; (2) developing talents into suitable and fulfilling careers that enhance interesting, rewarding lives; and (3) encouraging curiosity and intellectual risk-taking that might lead to original, useful ideas that propel a field—or even a civilization—forward.

But the biggest and saddest problem with the current state of gifted-and-talented education in our society is this: Support for the top end of the talent curve is still a privilege enjoyed primarily by students who are also at the top end of the socioeconomic curve. The gates that guard the path from talented youth to high-achieving adult remain essentially closed to poor and even many middle-class kids and to kids who grow up in places—such as southern Arkansas—where education is not a priority. Given that giftedness can spring from anywhere—from the blue blood of Boston's Back Bay or from a Coca-Cola bottler and a yoga instructor in Texarkana—it's worth pondering how many Taylors we are missing and the implications of that for our society.

"What our recent studies underscore is the tremendous amount of potential out there," says David Lubinski, the Vanderbilt University psychologist who codirects the Study of Mathematically Precocious Youth. "Our concern is that we're not coming close to identifying all of this population, that many of the smartest kids in the country thus don't get the special attention they need to explore and extend their interests, and develop their capabilities to a high level."

Even for those who are identified, the high price of programming deters many. The cost for the residential Johns Hopkins Center for Talented Youth's summer program ranges between $4,065 and $5,145 (although limited need-based scholarships are available), and Stanford's Education Program for Gifted Youth costs close to $17,000 a year—for an online program. Fees for the batteries of assessments and portfolio submissions required for admission can run into the thousands of dollars. Davidson Academy is tuition-free, but Colleen Harsin has no illusions about her school's failure to serve deserving children of limited means. "It's just not realistic," Harsin says, "for most people to do what Taylor's family did, quit their jobs and move across the country."

Research shows that genetic and environmental advantages multiply with the passage of years, as do disadvantages. Someone who was identified as gifted at thirteen and, as a result, given advanced instruction will be in a dramatically higher intellectual and socioeconomic position when he graduates from college than a similarly gifted peer who wasn't identified (or who was identified but for one reason or another did not receive appropriate support). In contrast to the well-endowed youth who leapfrog their classmates and convert high hopes into real achievement, gifted kids who are underchallenged typically become bored and frustrated, and their motivation and shoulders sink ever further. Thus, doors that have been shut for generations remain shut, and the gap between high-potential high achievers and high-potential low achievers grows wider.

There are bright spots, such as the Jack Kent Cooke Foundation's Young Scholars Program, which provides exceptionally promising, financially needy eighth graders with summer academic enrichment, study-abroad opportunities, and art or music master classes followed up with financial support and ongoing counseling and advocacy. And yet, as of 2014, the Young Scholars Program has been able to serve a total of only six hundred students.

David Feldman, the Tufts University child-prodigy researcher, has noted that it takes an uncanny convergence of circumstances lining up just right for talent to flourish. The consequence is that of the millions of children born with the potential to propel themselves to mastery, only a tiny portion are given the chance to become eminent, creative, high-achieving adults. Taylor was one of the lucky few. He was blessed with brains and curiosity and the gift of gab, with extraordinary parents who went to great lengths to ensure that he had access to an extraordinary education, and with a seemingly magic combination of other personal characteristics that contributed to his success. He drew in supportive mentors and teachers who believed in him and helped him realize his improbable dreams.

"That," says Feldman, "is a lot to get right."

The fact that the particular circumstances that helped Taylor develop his gifts have now largely evaporated demonstrates just how tenuous and capricious the whole talent-development process can be. The magic circles that Taylor fell into and that allowed him to become the Taylor he is today are mostly gone. Davidson Academy, though it is still the gold standard for developing certain kinds of talent, has drifted away from its freeform approach to interest-based learning. Declining education funding has made mentors at UNR ever more scarce. Ron Phaneuf is retired, and other professors have been furloughed or distracted by shrinking research budgets. "All of this makes it harder for the university to be a good partner with Davidson, to provide that support," says Bruno Bauer. "For example, there will be no new Bill [Brinsmead] coming on board. That position won't be replaced."

When it comes to supporting children who have the potential to become tomorrow's innovators, says Simonton, "it takes not just a village or school system but a whole sociocultural system."

Whether we use it or not, we have the recipe to build that system: parents who are courageous enough to give their children wings and let them fly in the directions they choose; schools that support children as individuals; a society that understands the difference between elitism and individualized education and that addresses the needs of kids at all levels.

Feldman may be right, for now at least, that the success of a gifted child depends largely on luck. But it doesn't have to be that way. As a democratic society, we do, to a large extent, make our own luck. We create the conditions that allow us—and our children, and the society of the future—to thrive, or to falter.

ACKNOWLEDGMENTS

I EXTEND MY HEARTFELT thanks to the Wilson family, whose openness, generosity, and trust made this book possible. Tiffany, Kenneth, Taylor, Joey, and Ashlee shared their home and lives and feelings with me and encouraged me to form my own impressions about their experiences. As result, I've been able to share the story of a real family whose journey gives us much to learn from.

Taylor's inner circle — especially Bill Brinsmead, Ron Phaneuf, and Carl Willis — generously and candidly opened up their laboratories, homes, and memories. Other friends and mentors who shared recollections and insights include Sofia Baig, David Boudreaux, Anthony Fidaleo, Ikya Kandula, Angela Melde, Dee Miller, George Ochs, Ellen Orr, Tristan Rasmussen, and Stephen Younger. I also appreciate the time and access provided by the faculty and staff at Davidson Academy, including Jan and Bob Davidson, Julie Dudley, Colleen Harsin, Melissa Lance, Darren Ripley, Alanna Simmons, and Elizabeth Walenta. At the University of Nevada–Reno, I'm grateful for the input of Benny Bach, Bruno Bauer, Wade Cline, Andrew Oxner, and Friedwardt Winterberg.

The Boy Who Played with Fusion had a crooked path from concept to completion — and as a result benefitted from the guidance of several talented editors, including *GQ*'s Mike Benoist and Donovan Hohn. The story migrated a few blocks downtown to *Popular Science,* helmed at the time by Mark Jannot, with whom I've had the longest and most productive writer/editor partnership of my career. While Mark encouraged me, as he always has, to make the story better and better, editor Cliff Ransom came up with a title that perfectly fit both magazine story and book.

Amanda Cook at Crown was the first to ask if I'd consider writing a book, and I greatly appreciate her interest as well as our initial exchange of ideas. The project eventually landed in the hands of Eamon Dolan at HMH, whose sharp instincts helped to transform a swollen mass of words into something resembling an artful narrative. Among the highlights of our collaboration were a series of brain-sampling sessions that rank among the most stimulating conversations I've ever had. These exchanges helped to shape the book's structure and give it thematic legs (as well as amputate a few extra limbs that threatened to trip it up).

Eamon's team managed the makeover from manuscript to book. Among those who did double duty to make the pages stand up straight were production editor Lisa Glover, copyeditor Tracy Roe, page designer Chrissy Kurpeski, and editorial assistants Ben Hyman and Rosemary McGuinness. I also appreciate the work of independent researchers Rachel Greene, Sarah Kuljian, Katherine Plumhoff, and Kenny Wassus.

I was lucky to have literary agent David McCormick on my side. Across the continent, my thanks go out to film agent Dana Spector at Paradigm Talent Agency. And across the Atlantic, my appreciation extends to Faber & Faber's Julian Loose, who brought Taylor's story to readers in the UK and the wider English-speaking world.

Some of my favorite writers and literary friends—Jeremiah Chamberlain, Bill Lychack, Miles Harvey, and Bob Parks—pulled themselves away from their own work to help me unearth the bits of clarity within my drafts, which benefitted greatly from their thoughtful and inspired ideas. Each of them saved me from literary embarrassment, just as astrophysicist Greg Tarlé and writer/physicist/magician Alex Stone saved me from scientific embarrassment. (Any errors that remain are, of course, my own.)

My gratitude extends to the scores of scientists, educators, psychologists, and others who shared knowledge, insights, and access. They include Susan Assouline, Sarah Andrew-Vaughan, Linda Brody, Shawn Carlson, Francisco Xavier Castellanos, Mark Cendrowski, Tom Chesshir, Jane Clarenbach, Caren Cooper, Steven Cowley, Chris Critch, Lee Ann Dickinson-Kelley, Carol Dweck, Lee Dodds, Anthony Fauci, David Henry Feldman, David Hahn, Richard Hull, Kristina Johnson, Ralf Kaiser, Scott Barry Kaufman, Felice Kaufmann, Barbara Kerr, Tom Ligon, David Lubinski, Dona Matthews, Edward Moses, Thiago Olson, Thomas Ruth, Joanne Ruthsatz, David Saltzberg, Ted Selker, Dean Keith Simonton, Martin Storksdieck, Rena Subotnik, Jonathan Wai, Ed Wingate, Ellen Winner, and Chuck Yu. If I've forgotten anyone, please accept my apology.

I've been gifted with a support system of generous, encouraging, and patient friends and family. My sisters—Julie, Melinda, and Karen—have always been my best friends, sustaining me with love and laughs. My parents have supported me in ways that include, most profoundly, an acceptance of the paths I've taken in my life, which are so very different from their own. The many friends who lent a physical or emotional hand include Jim Burnstein, Dede Cummings, Jennifer Jay, Anne Latchis, Laura Monschau, Dawn MacKeen, Cathy Mizgerd, Birgit Rieck, the Poplin family, Johnny Swing, Patrick Symmes, Pam Vitaz, Kimberly Rankin, Robin Westen, and Rachel Rotger—who emerged, late in the game, as a partner-in-crime and muse-in-chief.

Finally, I turn to my young sons, Charlie and Joe, to whom this book—and all else in my life—is dedicated. It's ironic that the process of writing a book that is largely about parenting would so severely diminish my own capacity to parent (so much so that I'm fairly sure I wouldn't do it again). Unique among those mentioned here, my children didn't choose to participate in this process—or to deal with an often distracted, sleep-deprived, or absent dad.

In some ways, though, the experience brought us closer together, and led us to a few beautiful and intriguing discoveries. Among them is the notion that our children can (and should) be our mentors. Charlie and Joe have taught me, most profoundly, that enthusiasm and raw curiosity can cut through a great many tough and challenging things. Now that my boys are growing older I can have conversations with them about the subjects I write about—or I can snap my laptop shut and roll off my chair to roughhouse with my rowdy little scientist/philosophers. Either way, what becomes clear is that playfulness and enthusiasm aren't just for kids. They are also ours, as adults, to reclaim and keep—if we choose to do so. Without my sons, I'd never have fully understood this.

As I rewrite the last sentences in this book for the last time, I fetch optimism from the realization that my sons, who were curious little boys when this project began, have grown into still-curious bigger boys; each a genius in his own way.

NOTES

INTRODUCTION

Page

xiv *identified as top performers by their teenage years:* Jonathan Wai, "Of Brainiacs and Billionaires," *Psychology Today,* 92 (2012): 78–85.

xv *personal attributes to shine:* Scott Barry Kaufman, "What Is Talent—and Can Science Spot What We Will Be Best At?," *Guardian,* July 6, 2013, http://www.theguardian.com/science/2013/jul/07/can-science-spot-talent-kaufman.

2. THE PRE-NUCLEAR FAMILY

11 *Hence Wilder filled and capped the first bottle of Coke:* "History," Coca-Cola/Dr Pepper Bottling Co., Nashville, Arkansas, http://nashvillecoca-coladrpepper.com/history.

4. SPACE CAMP

23 *orbiter rocket:* Daniel Lang, "A Romantic Urge," *New Yorker,* April 21, 1951, 183.

24 *Saturn project the go-ahead:* Matthew Brzezinski, *Red Moon Rising: Sputnik and the Hidden Rivalries That Ignited the Space Age* (New York: Times Books, 2007), 84–85, 87, 91–92.

according to the Marist Poll: Felicity Savage, "What Do You Want to Be When You Grow Up?," *Amazing Stories,* May 2, 2013.

top-ten career choice among children in the United Kingdom: Ibid.

26 *sports camps encouraged physical development:* Michael Neufeld, *Von Braun: Dreamer of Space, Engineer of War* (New York: Vintage, 2008), 354–55.

so-called STEM subjects: Karen Woodruff, "A History of STEM—Reigniting the Challenge with NGSS and CCSS," NASA Endeavor Science Teaching Certificate Project, http://www.us-satellite.net/STEMblog/?p=31.

5. THE "RESPONSIBLE" RADIOACTIVE BOY SCOUT

28 *Arkansas Municipal Auditorium:* "Rock & Roll Highway 67 and the Arkansas Municipal Auditorium, Texarkana, Arkansas," City of Texarkana, Arkan-

sas Agenda, http://arkagenda.txkusa.org/2011/05162011/05162011agenda_html /item_9_05162011_clerk_rock%20n%20roll%20hwy%2067_tmac.pdf.

33 *"I could do cooler stuff at home"*: David Hahn, telephone interview with the author, February 14, 2014. Unless otherwise noted, all quotes attributed to David Hahn are from this interview.

"creating energy": Ken Silverstein, *The Radioactive Boy Scout* (New York: Villard, 2005).

6. THE COOKIE JAR

39 *Marie Curie coined the term* radioactivity: John M. Reynolds, *An Introduction to Applied and Environmental Geophysics* (Chichester, UK: John Wiley and Sons, 2011), 1934.

overturned established ideas in chemistry: L. Pearce Williams, "Curie, Pierre and Marie," *Encyclopedia Americana,* vol. 8 (Danbury, CT: Grolier, 1986), 332.

40 *had little sense of the damage that radiation could do:* "Marie Curie: The Radium Institute (1919–1934)," American Institute of Physics, http://www.aip.org/history /curie/radinst3.htm.

worked unprotected for decades: Denise Grady, "A Glow in the Dark, and a Lesson in Scientific Peril," *New York Times,* October 6, 1998.

carried radioisotopes in her pocket: James Shipman, Jerry D. Wilson, and Aaron Todd, *An Introduction to Physical Science* (Boston: Houghton Mifflin, 2009), 257.

"One of our joys": Marie Curie, *Pierre Curie: With Autobiographical Notes,* trans. Charlotte and Vernon Kellogg (New York: Macmillan, 1923), 186–87.

Curie developed mobile x-ray stations: "Marie Curie: War Duty (1914–1919)," American Institute of Physics, http://www.aip.org/history/curie/war2.htm.

stored in locked lead-lined boxes: Craig J. Hogan, "We Are Made of Starstuff," *Science* 292 (May 2001): 863.

42 *than an equal number of atoms of uranium:* "Radium: The Benchmark," *Los Alamos Science* 23 (1995): 224–33.

permanent internal radiation source: Ibid.

7. IN THE (GLOWING) FOOTSTEPS OF GIANTS

47 *Roentgen made on November 8, 1895:* Robert A. Novelline, *Squire's Fundamentals of Radiology* (Cambridge, MA: Harvard University Press, 1997), 1.

49 *a high-frequency electromagnetic wave:* "Henri Becquerel—Facts," NobelPrize .org, http://www.nobelprize.org/nobel_prizes/physics/laureates/1903/becquerel -facts.html.

50 *Uranium is produced by supernovae:* "The Cosmic Origins of Uranium," World Nuclear Association, http://www.world-nuclear.org/info/Nuclear-Fuel-Cycle /Uranium-Resources/The-Cosmic-Origins-of-Uranium/.

polonium and radium: "Marie Curie—Facts," NobelPrize.org, http://www.nob elprize.org/nobel_prizes/physics/laureates/1903/marie-curie-facts.html.

51 *"It was as if you had fired a 15-inch naval shell"*: James Rutherford, Gerald Holton, and Fletcher Watson, *The Project Physics Course* (New York: Holt, Rinehart, and Winston, 1971).

the splitting of the atom: "The Nucleus: Rutherford, 1911," Cambridge Physics, http://www.outreach.phy.cam.ac.uk/camphy/nucleus/nucleus_index.htm.

52 *sources for the radioactive wares Hahn wanted to purchase:* Silverstein, *Radioactive Boy Scout.*

53 *unauthorized possession of nuclear material:* Eric Yosomono, Dustin Koski, and Evan Symon, "The 6 Most Reckless Uses of Radioactive Material," *Cracked,* January 7, 2012, http://www.cracked.com/article_19607_the-6-most-reckless-uses-radioactive-material_p2.html#ixzz2xM5e1ClB; "Richard Handl, Swedish Man, Tried to Set Up Nuclear Reactor at Home," *Huffington Post,* August 3, 2011, http://www.huffingtonpost.com/2011/08/03/richard-handl-nuclear-reactor-home_n_917585.html; Anthony Watts, "Don't Try Nuclear Energy Experiments at Home," WUWT, August 4, 2011, http://wattsupwiththat.com/2011/08/04/dont-try-nuclear-energy-experiments-at-home/.

the owner is required to notify the NRC: See http://www.nrc.gov/reading-rm/doc-collections/cfr/part031/part031-0012.html.

8. ALPHA, BETA, GAMMA

54 *radioactive gas in well water:* J. J. Thomson, "Radio-Active Gas from Well Water," *Nature* 67, no. 1748 (1903): 609.

surgeon general George H. Torney: Paul W. Frame, "Radioactive Curative Devices and Spas," Oak Ridge Associated Universities, November 5, 1989, http://www.orau.org/ptp/articlesstories/quackstory.htm.

"radioactivity prevents insanity": Ibid.

buried in a lead-lined coffin: Charlie Hintz, "The Radioactive Death of Eben Byers," Cult of Weird, December 15, 2010, http://www.cultofweird.com/medical/eben-byers-radithor-poisoning/.

57 *28 from acute radiation exposure:* "Health Impacts: Chernobyl Accident Appendix 2," World Nuclear Association, last modified November 2009, http://www.world-nuclear.org/info/Safety-and-Security/Safety-of-Plants/Appendices/Chernobyl-Accident---Appendix-2--Health-Impacts/.

the number of eventual casualties caused by the radiation plume: United Nations Chernobyl Forum, "Health Effects of the Chernobyl Accident and Special Health Care Programmes," World Health Organization, last modified 2006, http://www.who.int/ionizing_radiation/chernobyl/WHO%20Report%20on%20Chernobyl%20Health%20Effects%20July%2006.pdf.

four thousand additional cancer deaths: "Chernobyl: The True Scale of the Accident," World Health Organization, last modified September 5, 2005, http://www.who.int/mediacentre/news/releases/2005/pr38/en/.

International Agency for Research on Cancer: "The Cancer Burden from Chernobyl in Europe," International Agency for Research on Cancer, last modified April 20, 2006, http://www.iarc.fr/en/media-centre/pr/2006/pr168.html.

invisible particles they breathed: Tom Zoellner, *Uranium: War, Energy, and the Rock That Shaped the World* (New York: Viking, 2010), 172–73.

58 *smear their reputations:* Ross Mullner, *Deadly Glow: The Radium Dial Worker Tragedy* (Washington, DC: American Public Health Association, 1999); William Kovarik, "The Radium Girls," updated from chapter 8 in Mark Neuzil and William Kovarik, *Mass Media and Environmental Conflict: America's Green Crusades* (Thousand Oaks, CA: Sage Publications, 1996); Denise Grady, "A Glow in the Dark, and a Lesson in Scientific Peril," *New York Times,* October 6, 1998.

59 *"you can guarantee it will be closed down":* Carl Willis, telephone interview with the author, March 20, 2014.

"like some of the old Third World medical machines": William Kolb, interview with the author, October 5, 2013, Richmond, Virginia.

61 *facility northeast of Tokyo:* NHK TV crew, *A Slow Death: 83 Days of Radiation Sickness* (New York: Vertical, 2008).

9. TRUST BUT VERIFY

66 *"don't call it transmutation":* Muriel Howorth, *Pioneer Research on the Atom: The Life Story of Frederick Soddy* (London: New World, 1958), 83–84; Lawrence Badash, "Radium, Radioactivity, and the Popularity of Scientific Discovery," *Proceedings of the American Philosophical Society* 122, no. 3 (June 9, 1978): 145–54; Thaddeus J. Trenn, *The Self-Splitting Atom: The History of the Rutherford-Soddy Collaboration* (London: Taylor and Francis, 1977), 42, 58–60, 111–17.

10. EXTREME PARENTING

72 *"steering their children toward things the parents have chosen":* Ellen Winner, telephone interview with the author, January 14, 2014. Unless otherwise noted, all quotes attributed to Ellen Winner are from this interview.

"that's the hardest one to get right": David Henry Feldman, telephone interview with the author, January 14, 2014. Unless otherwise noted, all quotes attributed to David Henry Feldman are from this interview.

74 *"notice what they're picking up on":* Linda Brody, telephone interview with the author, January 21, 2014. Unless otherwise noted, all quotes attributed to Linda Brody are from this interview.

75 *"a lot of different ways of looking at different things":* Dean Keith Simonton, quoted in Ann Hulbert, "The Prodigy Puzzle," *New York Times,* November 20, 2005.

new psychological research suggests: Laurence Steinberg, *Age of Opportunity: Lessons from the New Science of Adolescence* (Boston: Houghton Mifflin Harcourt, 2014).

11. ACCELERATING TOWARD BIG SCIENCE

80 *"There are critical times":* Barbara A. Kerr, ed., *Encyclopedia of Giftedness, Creativity, and Talent* (Thousand Oaks, CA: Sage Publications, 2009).

"it does a bright child no favors": Barbara Kerr, telephone interview with the author, January 22, 2014. Unless otherwise noted, all quotes attributed to Barbara Kerr are from this interview.

trained in identifying and supporting gifted and talented students: Editorial Board, "Even Gifted Students Can't Keep Up," *New York Times,* December 14, 2013.

achieved higher levels of mastery as adults: Rena F. Subotnik, Paula Olszewski-Kubilius, and Frank C. Worrell, "Rethinking Giftedness and Gifted Education: A Proposed Direction Forward Based on Psychological Science," *Psychological Science in the Public Interest* 12, no. 1 (2011): 3–54.

81 *one or both parents in a scientific field:* Rena F. Subotnik et al., "Specialized Public High Schools of Science, Mathematics, and Technology and the STEM Pipeline: What Do We Know Now and What Will We Know in 5 Years?," *Roeper Review* 32, no. 1 (2009): 7–16.

83 *"Sons, obey your fathers"*: Homer Hickam, *Rocket Boys* (New York: Dell, 1999), 104–6.

12. HEAVY WATER

89 *"work hard to learn more and get smarter"*: Carol Dweck, telephone interview with the author, January 27, 2014. Unless otherwise noted, all quotes attributed to Carol Dweck are from this interview.

poll of 143 creativity researchers: Robert J. Sternberg, *Handbook of Creativity* (New York: Cambridge University Press, 1998).

14. BRINGING THE STARS DOWN TO EARTH

103 *"and preferably a lot sooner"*: Steven Cowley, telephone interview with the author, November 8, 2012. Unless otherwise noted, all quotes attributed to Steven Cowley are from this interview.

106 *"making an Apollo-like commitment?"*: Ralf Kaiser, interview with the author, San Diego, October 12, 2012. Unless otherwise noted, all quotes attributed to Ralf Kaiser are from this interview.

107 *"Positive images of the future"*: E. Paul Torrance, "The Importance of Falling in Love with 'Something,'" *Creative Child and Adult Quarterly* 8 (1983): 72–78.

boost both cognitive efficiency and overall productivity: Harry Alder, *CQ: Boost Your Creative Intelligence: Powerful Ways to Improve Your Creativity Quotient* (Philadelphia: Kogan Page, 2002); Jing Zho and Christina E. Shalley, "Research on Employee Creativity: A Critical Review and Directions for Further Research," *Research in Personnel and Human Resources Management* 22 (2003): 165–217.

108 *"went down a road a more experienced person wouldn't have"*: Anthony Fauci, telephone interview with the author, January 5, 2015. Unless otherwise noted, all quotes attributed to Anthony Fauci are from this interview.

15. ROOTS OF PRODIGIOUSNESS

110 *Definitions of giftedness vary:* Joan Freeman, "Teaching Gifted Pupils," *Journal of Biological Education* 33 (1999): 185–90.

between autism and prodigiousness: Joanne Ruthsatz and Jourdan B. Urbach, "Child Prodigy: A Novel Cognitive Profile Places Elevated General Intelligence, Exceptional Working Memory, and Attention to Detail at the Root of Prodigiousness," *Intelligence* 40 (2012): 419–26.

111 *"significantly ahead of their year group":* UK Department for Children, Schools, and Families, *Identifying Gifted and Talented Learners—Getting Started* (Nottingham, UK: DCSF Publications, 2008), 1.

35 to 10 percent: Louise Wheeler, "Is Your Child Super Smart?," *SheKnows,* December 20, 2010, http://www.sheknows.co.uk/parenting/articles/821475/signs-your-child-is-gifted.

Turkey: Erisa Dautaj Şenerdem, "Turkish Ministry Examines Education Options for Gifted Children," *Hürriyet Daily News,* January 18, 2011.

India: Summiya Yasmeen, "Wasted Potential of India's Gifted Children," *Education World,* http://www.educationworldonline.net/index.php/page-article-choice-more-id-862.

112 *prodigy has a distinct form of giftedness:* Ellen Winner, "The Miseducation of Our Gifted Children," *Education Week,* October 16, 1996, http://www.edweek.org/ew/articles/1996/10/16/07winner.h16.html.

113 *neuroscientist Francisco Xavier Castellanos:* Francisco Xavier Castellanos, telephone interviews with the author, February 8, 2014, and December 29, 2014. Unless otherwise noted, all quotes attributed to Francisco Xavier Castellanos are from these interviews.

responses to pharmacological or behavioral treatments: See http://www.cell.com/neuron/abstract/S0896-6273(14)00967-2.

gray matter that makes up the brain's outer layer: P. Shaw et al., "Intellectual Ability and Cortical Development in Children and Adolescents," *Nature* 440 (2006): 676–79.

between the parietal and frontal lobes: Ker Than, "New Theory: How Intelligence Works," *LiveScience,* September 11, 2007, http://www.livescience.com/1863-theory-intelligence-works.html.

114 *average in the 140 range:* Ruthsatz and Urbach, "Child Prodigy."

"frightening ease": James Gleick, *Genius: The Life and Science of Richard Feynman* (New York: Vintage), 129.

hold information in our minds in a highly active state: Randall W. Engle, "Working Memory Capacity as Executive Attention," *Current Directions in Psychological Science* 11 (2002): 19–23.

99th percentile for working memory: Ruthsatz and Urbach, "Child Prodigy"; Elizabeth J. Meinz and David Z. Hambrick, "Deliberate Practice Is Necessary but Not Sufficient to Explain Individual Differences in Piano Sight-Reading Skill: The Role of Working Memory Capacity," *Psychological Science* 21, no. 7 (2010): 914–19; Joanne Ruthsatz and Douglas K. Detterman, "An Extraordinary Memory: The Case Study of a Musical Prodigy," *Intelligence* 31, no. 6 (2003): 509–18.

115 *no solid scientific evidence to back up these promises:* See http://longevity3.stanford.edu/blog/2014/10/15/the-consensus-on-the-brain-training-industry-from-the-scientific-community/.

working-memory capacities: Emily Finn, "When Four Is Not Four, but Rather Two Plus Two," *MIT News*, June 23, 2011, http://newsoffice.mit.edu/2011/miller-memory-0623.

reinvent modern culture: Harrison J. Kell, David Lubinski, and Camilla P. Benbow, "Who Rises to the Top? Early Indicators," *Psychological Science* 24 (2013): 648–59.

116 *Study of Mathematically Precocious Youth:* D. Lubinski and C. P. Benbow, "Study of Mathematically Precocious Youth After 35 years: Uncovering Antecedents for the Development of Math-Science Expertise," *Perspectives on Psychological Science* 1 (2006): 316–45.

to a great extent heritable: Than, "New Theory"; Jonathan Wai, "Experts Are Born, Then Made: Combining Prospective and Retrospective Longitudinal Data Shows That Cognitive Ability Matters," *Intelligence* 45 (2014): 74–80.

117 *than on innate ability or talent:* K. Anders Ericsson, Ralf T. Krampe, and Clemens Tesch-Römer, "The Role of Deliberate Practice in the Acquisition of Expert Performance," *Psychological Review* 100 (1993): 363–406.

"differences in the recorded amounts of deliberate practice": Ibid.

"experts are always made, not born": K. Anders Ericsson, Michael J. Prietula, and Edward T. Cokely, "The Making of an Expert," *Harvard Business Review* (2007): 1–8.

The duffer, who started at 30 over par, told the BBC: Ben Carter, *BBC News*, February 28, 2014, http://www.bbc.com/news/magazine-26384712.

118 *highly related to cognitive ability:* Wai, "Experts Are Born, Then Made."

practice time explains only 20 to 25 percent of performance differences: David Z. Hambrick et al., "Deliberate Practice: Is That All It Takes to Become an Expert?," *Intelligence* 45 (2014): 34–45; Brooke N. Macnamara, David Z. Hambrick, and Frederick L. Oswald, "Deliberate Practice and Performance in Music, Games, Sports, Education, and Professions: A Meta-Analysis," *Psychological Science* 25, no. 8 (2014): 1608–18.

has led several studies: Dean Keith Simonton, "Career Landmarks in Science: Individual Differences and Interdisciplinary Contrasts," *Developmental Psychology* 27 (1991): 119–30; Dean Keith Simonton, "Leaders of American Psychology, 1879–1967: Career Development, Creative Output, and Professional Achievement," *Journal of Personality and Social Psychology* 62 (1992): 5–17; Dean Keith Simonton, "Creative Productivity: A Predictive and Explanatory Model of Career Trajectories and Landmarks," *Psychological Review* 104 (1997): 66–89; Dean Keith Simonton, "Talent and Its Development: An Emergenic and Epigenetic Model," *Psychological Review* 106 (1999): 435–57.

"no such thing as innate talent": Scott Barry Kaufman, "Talent Versus Practice," *Beautiful Minds*, July 15, 2014, http://blogs.scientificamerican.com/beautiful-minds/2014/07/15/talent-vs-practice-why-are-we-still-debating-this-anymore/.

119 *"why exceptional talent is so rare"*: Dean Keith Simonton, interview with the author, July 17, 2014. Unless otherwise noted, other quotes attributed to Dean Keith Simonton are from this interview.

personal attributes that lead to extraordinary performance: David Lubinski, telephone interviews with the author, July 15, 2014, and December 30, 2014. Unless otherwise noted, all quotes attributed to David Lubinski are from these interviews.

120 *smart peers who didn't have these opportunities:* Gregory Park, David Lubinski, and Camilla P. Benbow, "When Less Is More: Effects of Grade Skipping on Adult STEM Accomplishments Among Mathematically Precocious Youth," *Journal of Educational Psychology* 105, no. 1 (2013): 176–98; Jonathan Wai et al., "Accomplishment in Science, Technology, Engineering, and Mathematics (STEM) and Its Relation to STEM Educational Dose: A 25-Year Longitudinal Study," *Journal of Educational Psychology* 102, no. 4 (2010): 860–71.

16. THE LUCKY DONKEY THEORY

122 *"Bussard is a fictional character"*: Tom Ligon, telephone interview with the author, July 29, 2010. Unless otherwise noted, all quotes attributed to Tom Ligon are from this interview.

mainline fusion program: Tom Ligon, "The World's Simplest Fusion Reactor Revisited," *Analog Science Fiction and Fact,* January–February 2008.

"Einstein's famous formula": Ibid.

123 *exponential efficiency improvements:* Robert W. Bussard, "Inertial Electrostatic Fusion Systems Can Now Be Built," Fusor.net forums, accessed February 22, 2007.

124 *Jamie had not decisively achieved fusion:* See http://www.fusor.net/board/viewtopic.php?f=2&t=9292.

126 Inside Steve's Brain*:* Leander Kahney, *Inside Steve's Brain* (New York: Portfolio, 2009).

blowing things up with chemistry sets: Steve Silberman, "Don't Try This at Home," *Wired,* June 2006.

"stinks and bangs and crystals and colors": Ibid.

127 *National Science Board:* "Higher Education in Science and Engineering," National Science Foundation. http://www.nsf.gov/statistics/seind14/index.cfm/chapter-2.

128 *"with or without an Erlenmeyer flask"*: Silberman, "Don't Try This at Home."

"explore the world with their hands": Ted Selker, telephone interview with the author, December 16, 2014. Unless otherwise noted, all quotes attributed to Ted Selker are from this interview.

130 *"big science has the special problem"*: Steven Weinberg "The Crisis of Big Science," *New York Review of Books,* May 10, 2012, http://www.nybooks.com/articles/archives/2012/may/10/crisis-big-science/?pagination=false.

"low-hanging fruits (like Newton's apple) have been plucked": George Johnson,

"Hills to Scientific Discoveries Grow Steeper," *New York Times,* February 18, 2014.

17. TWICE AS NICE, HALF AS GOOD

135 *"and stir up mischief"*: Susan Freinkel, "IQ Like Einstein," http://www.greatschools.org/print-view/parenting-dilemmas/7562-profoundly-gifted-child-story.gs.

136 *"reach upper primary or junior secondary school"*: See http://www.scmp.com/lifestyle/family-education/article/1412655/academy-helps-students-overcome-challenges-being-gifted.

137 *"industrialized model"*: Ken Robinson, "How Schools Kill Creativity," TED Talks, February 2006, http://www.ted.com/talks/ken_robinson_says_schools_kill_creativity?language=en.

one long-term study: Kell, Lubinski, and Benbow, "Who Rises to the Top?"

138 *as high as 5 percent:* Larisa Shavinina, ed., *International Handbook on Giftedness* (New York: Springer, 2009).

nearly half of gifted children are underachieving: Ellen Winner, "Exceptionally High Intelligence and Schooling," *American Psychologist* 52, no. 10 (1997): 1070–81.

139 *"reinforce that they belong there"*: See http://www.bostonglobe.com/ideas/2014/03/15/the-poor-neglected-gifted-child/rJpv8G40eawWBBvXVtZyFM/story.html.

140 *"gifted programs have suffered"*: John Cloud, "Are We Failing Our Geniuses?," *Time,* August 16, 2007.

141 *U.S. students compared to their global counterparts:* See http://www.baltimoresun.com/news/opinion/oped/bs-ed-javits-program-20140317,0,4576118.story#ixzz2xAKX8RAC.

U.S. performance had changed little: Motoko Rich, "American 15-Year-Olds Lag, Mainly in Math, on International Standardized Tests," *New York Times,* December 3, 2013.

outscored the United States impressively in math and science: Stephanie Simon, "PISA Results Show 'Educational Stagnation' in U.S.," *Politico,* December 3, 2013, http://www.politico.com/story/2013/12/education-international-test-results-100575.html#ixzz30JaivozY.

average or below-average children: Robert Theaker et al., "Do High Flyers Maintain Their Altitude? Performance Trends of Top Students," Thomas B. Fordham Institute, September 20, 2011, http://edexcellence.net/publications/high-flyers.html.

18. ATOMIC TRAVEL

147 *experiments affecting the final design of the bomb that leveled Nagasaki:* David Hawkins, Edith C. Truslow, and Ralph Carlisle Smith, *Manhattan District History, Project Y, the Los Alamos Story* (Los Angeles: Tomash, 1963), 203.

19. CHAMPIONS FOR THE GIFTED

153 *asked thirteen thousand kids in seven states:* June Kronholz, "Challenging the Gifted: Nuclear Chemistry and Sartre Draw the Best and Brightest to Reno," *Education Next,* March 22, 2011, http://educationnext.org/challenging-the-gifted.

21. A FOURTH STATE OF GRAPE

177 *"Talent doesn't make you gritty":* Angela Duckworth, "The Key to Success? Grit," TED Talks, April 2013, http://www.ted.com/talks/angela_lee_duckworth_the_key_to_success_grit?language=en.

178 *"If we can secure interest":* John Dewey, *Interest and Effort in Education* (Boston: Houghton Mifflin, 1913).

New studies have reinforced Dewey's theories: Paul A. O'Keefe and Lisa Linnenbrink-Garcia, "The Role of Interest in Optimizing Performance and Self-Regulation," *Journal of Experimental Social Psychology* 53 (2014): 70–78.

make learning more efficient and facilitate creativity: Todd M. Thrash et al., "Mediating Between the Muse and the Masses: Inspiration and the Actualization of Creative Ideas," *Journal of Personality and Social Psychology* 98, no. 3 (2010): 469–87.

boost one's fortitude to persevere: Hong Jiewen and Angela Y. Lee, "Be Fit and Be Strong: Mastering Self-Regulation Through Regulatory Fit," *Journal of Consumer Research* 34, no. 5 (2008): 682–95.

22. HEAVY METAL APRON

189 *"At the little-c level":* Karen Kersting, "What Exactly Is Creativity?," *American Psychological Association Monitor* (November 2003), http://www.apa.org/monitor/nov03/creativity.aspx.

"Beethoven's music": Norman R. Augustine, "Educating the Gifted," *Psychological Science in the Public Interest* 12, no. 1 (January 2011): 1–2.

23. BIRTH OF A STAR

198 *"Hey Guys":* Taylor Wilson, "Archived—Plasma," Fusor.net, March 5, 2009, http://www.fusor.net/board/viewtopic.php?f=18&t=7819&p=56500#p56500.

24. THE NEUTRON CLUB

210 *"like bringing a Ferrari to a go-cart race":* Judy Dutton, *Science Fair Season* (New York: Hyperion, 2011), 29.

25. A FIELD OF DREAMS, AN EPIPHANY IN A BOX

220 *"awash in 2,000 metric tons"*: Editorial Board, "Increased Security for Nuclear Materials," *New York Times,* January 10, 2014, http://www.nytimes.com/2014/01/11/opinion/increased-security-for-nuclear-materials.html?src=recg.

221 *"another type of radioactive dispersal device is used"*: George M. Moore, "If the Boston Marathon Attack Had Involved Dirty Bombs," *Bulletin of the Atomic Scientists,* May 1, 2013, http://thebulletin.org/if-boston-marathon-attack-had-involved-dirty-bombs.

223 *supplies are quickly running out:* Jon Cartwright, "Shortages Spur Race for Helium-3 Alternatives," *Chemistry World,* January 12, 2012, http://www.rsc.org/chemistryworld/News/2012/January/helium-3-isotopes-shortage-alternatives-neutron-detectors.asp.

26. THE FATHER OF ALL BOMBS

228 *"super-high-temperature fireball and a massive shock wave"*: Vladimir Isachenkov, "Russia Tests Powerful 'Dad of All Bombs,'" *Washington Post,* September 11, 2007.

232 *"a family dynamic that I see too often"*: Dona Matthews, telephone interview with the author, December 20, 2014. Unless otherwise noted, all quotes attributed to Dona Matthews are from this interview.

"not just the superstar": Dona Matthews and Joanne Foster, *Beyond Intelligence: Secrets for Raising Happily Productive Kids* (Toronto: House of Anansi, 2014).

233 *"dominance-seeking behaviors"*: Susan Cain, *Quiet: The Power of Introverts in a World That Can't Stop Talking* (New York: Broadway Books, 2013).

27. WE'RE JUST BREATHING YOUR AIR

238 *"reminds some of the early Bill Gates"*: Eamonn Fingleton, "Is This the Bill Gates of Energy?" Forbes.com, November 26, 2012, http://www.forbes.com/sites/eamonnfingleton/2012/11/26/is-this-the-bill-gates-of-energy-meet-nuclear-entrepreneur-taylor-wilson-18/?ss=business:energy.

28. THE SUPER BOWL OF SCIENCE

248 *two-thirds of all nuclear medicine procedures:* National Research Council, *Medical Isotope Production Without Highly Enriched Uranium* (Washington, DC: National Academies Press, 2009), 68.

EPILOGUE

273 *some domain of expertise:* Kaufman, "What Is Talent."

278 *"promote exceptional youthful promise"*: Hulbert, "Prodigy Puzzle."

INDEX